Collins

AQA A-level
Chemistry

Sam Holyman

Organise
and **Retrieve** your
Knowledge

Acknowledgements

The author and publisher are grateful to the copyright holders for permission to use quoted materials and images.

Every effort has been made to trace copyright holders and obtain their permission for the use of copyright material. The author and publisher will gladly receive information enabling them to rectify any error or omission in subsequent editions.

All other images © Shutterstock.com or © HarperCollins*Publishers*

All facts are correct at time of going to press.

Published by Collins
An imprint of HarperCollins*Publishers* Limited
1 London Bridge Street
London SE1 9GF

HarperCollins*Publishers*
Macken House
39/40 Mayor Street Upper
Dublin 1
D01 C9W8
Ireland

British Library Cataloguing in Publication Data.

A CIP record of this book is available from the British Library.

Author: Samantha Holyman
Publisher: Clare Souza
Commissioning: Richard Toms
Project Management and editorial: Richard Toms and Shelley Teasdale
Inside concept design: Ian Wrigley
Layout: Ian Wrigley and Contentra Technologies
Cover design: Sarah Duxbury
Production: Bethany Brohm
Printed in the United Kingdom by Martins the Printers

MIX
Paper | Supporting responsible forestry
FSC™ C007454

This book contains FSC™ certified paper and other controlled sources to ensure responsible forest management.

For more information visit: www.harpercollins.co.uk/green

How to use this book

Each topic is presented on a two-page spread

Use the QR code to access a support video for the topic

Topics exclusive to A-level are labelled

Make your own list of key words for the topic

Answer the key questions to ensure you understand fundamental ideas

Organise your knowledge with concise explanations and examples

Summarise the key ideas using your own words or sketches

Embed your knowledge with retrieval questions for each topic (use separate sheets of paper if you need more space for workings)

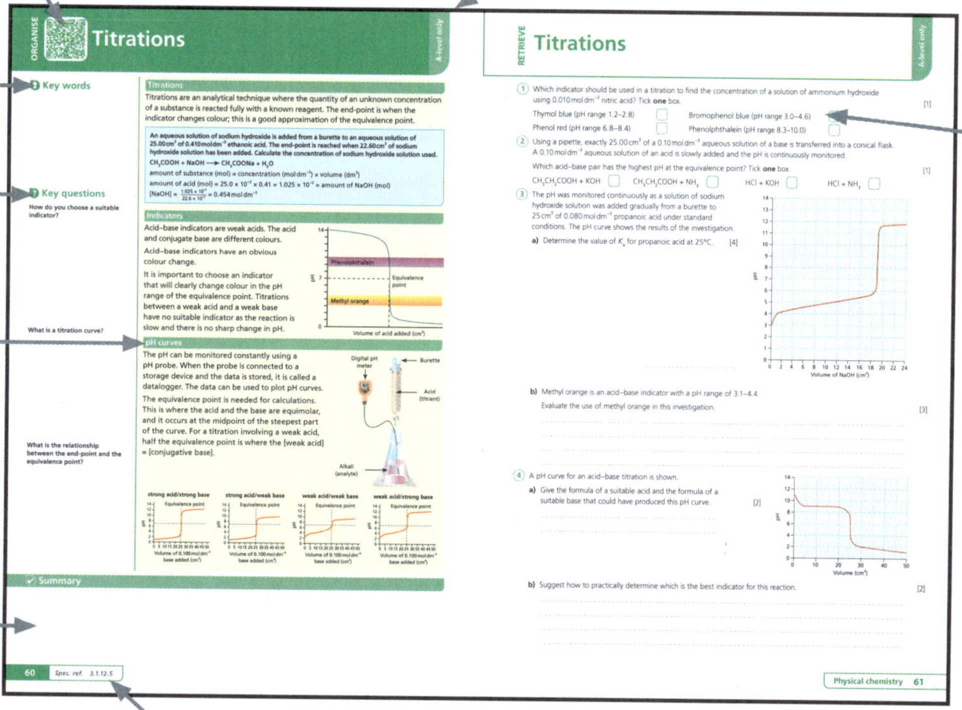

Specification references show you which part of the AQA course is being covered

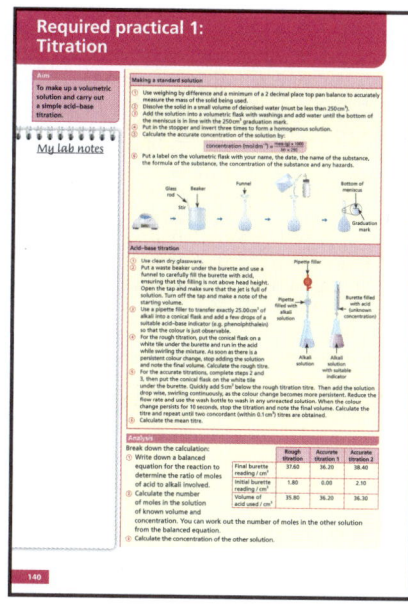

Explore each required practical and make your own supporting notes

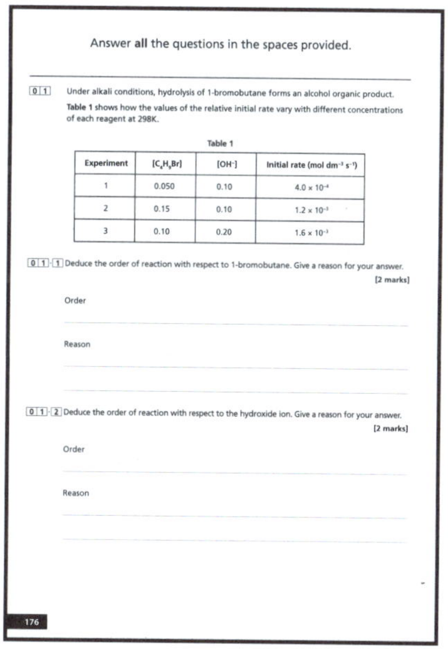

Try the practice papers to help prepare for the exams

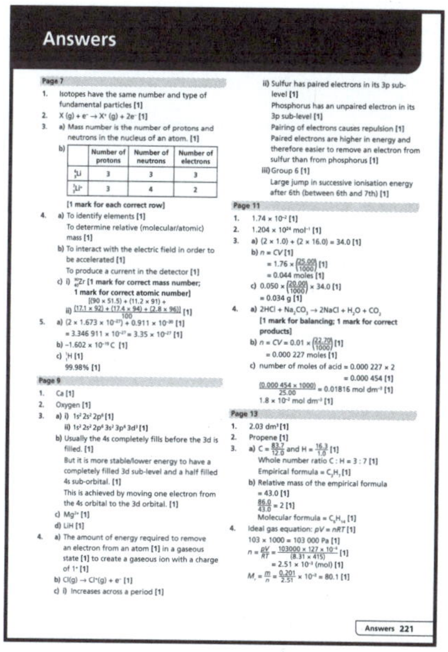

Answers are provided at the back of the book (excluding for the key questions, which aim to encourage self-study)

To access the ebook, visit **collinshub.co.uk/ebooks** and follow the step-by-step instructions.

Contents

Contents

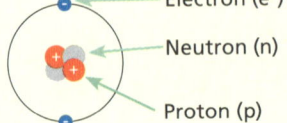

Fundamental particles

Key words

Key questions

How are electrons arranged in an atom?

Atomic structure

The current model of the atom has evolved out of technological advancements.

	Proton	Neutron	Electron
Mass (kg)	1.673×10^{-27}	1.673×10^{-27}	0.911×10^{-30}
Relative mass	1	1	0
Charge (C)	$+1.602 \times 10^{-19}$	0	-1.602×10^{-19}
Relative charge	+1	0	−1

Mass number (A)
Total number of neutrons and protons
Atomic (proton) number (Z)
Number of protons

$^4_2$He
Helium

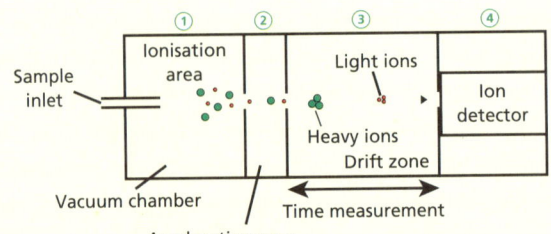

Electron (e⁻)
Neutron (n)
Proton (p)

Time of flight (TOF) mass spectrometry

Mass spectrometry can be used to identify elements, calculate their relative atomic masses, A_r, and determine the abundance of isotopes.

Isotopes of the same element have the same atomic number but a different mass number.

Simplified Diagram of a Time of Flight Spectrometer

① ② ③ ④

Sample inlet
Ionisation area
Light ions
Ion detector
Heavy ions
Drift zone
Vacuum chamber
Time measurement
Acceleration area

How do electrons fill the atomic orbitals?

① The particles in the sample are ionised to form 1⁺ ions by either electron spray or electron impact. The ions are accelerated with the same kinetic energy using an electric field.

② The ionised sample enters the flight tube through a small hole in a negatively charged plate.

③ The velocity of the ions is affected by their mass and the TOF is proportional to the square root of the mass of the ions. Lighter ions take less time to travel the flight tube and hit the detector plate first.

④ At the detector plate, the ions gain electrons and this results in an electric current. The greater the current, the more of a particular type of ion there is in the sample.

What is ionisation energy?

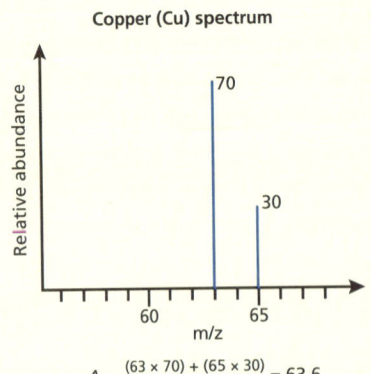

Copper (Cu) spectrum

$$A_r = \frac{(63 \times 70) + (65 \times 30)}{100} = 63.6$$

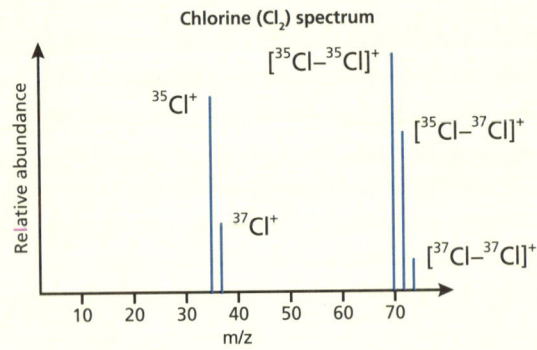

Chlorine (Cl_2) spectrum

Elements that form diatomic molecules give more complex spectra.

Summary

Fundamental particles

1 Which statement about isotopes of an element is **not** true? Tick **one** box. [1]

Isotopes have different physical properties ☐   Isotopes have the same chemical properties ☐

Isotopes have the same number and type of ☐   Isotopes have different mass numbers ☐
fundamental particles

2 In a time of flight mass spectrometer, atom X is ionised using electrospray ionisation.

What is the equation for this ionisation? Tick **one** box. [1]

$X (g) \rightarrow X^+ (g) + e^-$ ☐ $X (g) + e^- \rightarrow X^+ (g) + 2e^-$ ☐

$X (l) + e^- \rightarrow X^+ (g) + 2e^-$ ☐ $X (g) + H^+ \rightarrow XH^+ (g)$ ☐

3 a) Define the term 'mass number'. [1]

....

b) Complete the table to show the number of fundamental particles in two species of lithium. [2]

	Number of protons	Number of neutrons	Number of electrons
^6_3Li	3		
$^7_3Li^+$	3		

4 a) Give **two** uses of mass spectrometry. [2]

....

....

b) In mass spectrometry, explain why it is important that the sample is ionised. [2]

....

....

c) A mass spectrum is shown for the element zirconium.

i) Determine the formula for the most abundant species. [2]

....

ii) Show that the relative atomic mass, A_r, of zirconium is 91.3 [1]

....

5 The table at the top of page 6 shows some information about fundamental particles.

a) Calculate the mass of an isotope of hydrogen that has the formula ^2_1H. [2]

....

b) A different isotope of hydrogen forms a hydride ion with the formula $^3_1H^-$.

Determine the charge of this ion. .... [1]

c) The relative formula mass for hydrogen is 1.00. In a natural sample of hydrogen, there is about 0.02% ^2_1H and trace amounts of ^3_1H.

Give the formula and abundance of the most common isotope of hydrogen. [2]

....

....

Electron configuration

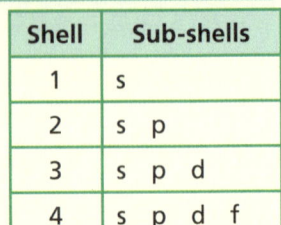

Key words

Key questions

How are electrons arranged in an atom?

How do electrons fill the atomic orbitals?

How do you write electronic structures?

What is ionisation energy?

Energy shells and sub-shells

The maximum number of electrons in a shell is $2n^2$, where n is the main number of the electron shell. The electron cloud model of the atom has been developed to show the space around the nucleus where the electrons are likely to be found in sub-shells and orbitals.

s sub-shells hold 1 orbital, p sub-shells hold 3 orbitals, d sub-shells hold 5 orbitals and f sub-shells hold 7 orbitals.

The first ionisation energy of Period 3 (and Group 2) is evidence of the electron structures:

Shell	Sub-shells
1	s
2	s p
3	s p d
4	s p d f

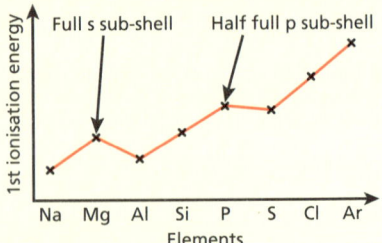

① This 1 electron is in the 3rd shell, furthest from the nucleus, experiencing comparatively weak attractive forces from the nucleus.

② These 2 electrons are in the 1st shell, closest to the nucleus, experiencing very strong attractive forces from the nucleus.

③ These 8 electrons are in the 2nd shell, experiencing reasonably strong attractive forces from the nucleus.

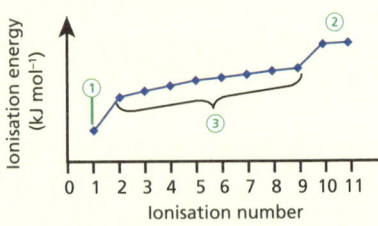

Electron configuration of atoms and ions

Electrons enter the lowest energy orbital available. They fill one electron per orbital for each sub-shell before a second electron enters the orbital.

The summarised electronic structure for iron is: $1s^2\ 2s^2\ 2p^6\ 3s^2\ 3p^6\ 3d^6\ 4s^2$

Electron configuration can be abbreviated by using the symbol for a noble gas to represent its configuration and then listing the remaining electrons.

The abbreviated electron structure for iron is [Ar] $3d^6\ 4s^2$

Energy level diagram for electrons in an iron atom (each box represents an orbital holding up to two electrons with opposite spin)

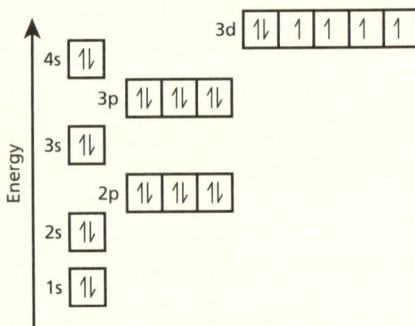

Metal atoms lose their outer-shell electrons to make positive ions so their electron structure has fewer electrons than the atom of the element.

Iron(II), $Fe^{2+} = 1s^2\ 2s^2\ 2p^6\ 3s^2\ 3p^6\ 3d^6 = $ [Ar] $3d^6$

Iron(III), $Fe^{3+} = 1s^2\ 2s^2\ 2p^6\ 3s^2\ 3p^6\ 3d^5 = $ [Ar] $3d^5$

Non-metal atoms gain electrons in their outer shell to become negative ions so their electron structure has more electrons than the atom of the element.

Oxide, $O^{2-} = 1s^2\ 2s^2\ 2p^6$ or [He] $2s^2\ 2p^6$

Chloride, $Cl^- = 1s^2\ 2s^2\ 2p^6\ 3s^2\ 3p^6$ or [Ne] $3s^2\ 3p^6$

✔ Summary

Electron configuration

1 Which species does **not** have the electronic structure $1s^2\ 2s^2\ 2p^6\ 3s^2\ 3p^6$? Tick **one** box. [1]

Ca ☐ Ca^{2+} ☐   Ar ☐ S^{2-} ☐

2 Which atom contains only two unpaired electrons? Tick **one** box. [1]

Oxygen ☐   Calcium ☐   Boron ☐   Magnesium ☐

3 **a)** Write the full electron configuration for the following species. [2]

     i) F^- ....   **ii)** V^{2+} ....

b) Copper has the electron configuration $[Ar]\ 4s^1\ 3d^{10}$

Explain why the electron configuration of copper is unusual. [3]

....

....

c) Deduce the formula of an ion that has a 2^+ charge and the same electron configuration as neon. [1]

....

d) Deduce the formula of the compound that contains 1^+ ions and 1^- ions that both have the same electron configuration as helium. [1]

....

4 **a)** Define the term 'first ionisation energy'. [3]

....

....

....

b) Write an equation to represent the process that occurs when the first ionisation energy for chlorine is measured. [1]

....

c) The graph shows the first ionisation energies for Period 3.

     i) Describe the general trend in first ionisation energies as you go across Period 3. [1]

....

....

ii) Explain why the first ionisation energy for sulfur is lower than phosphorus. [4]

....

....

....

....

iii) The table shows the successive ionisation energies for an atom.

Ionisation number	1	2	3	4	5	6	7	8
Ionisation energy (KJ mol^{-1})	1000	2260	3390	4540	6990	8490	27 100	31 700

Deduce the group number for this element. Explain your answer using data from the table. [2]

....

....

The mole

Key words

Key questions

What is relative atomic mass?

How do you calculate relative formula mass?

What is significant about Avogadro's number?

How do you calculate the amount of substance for the mass of a known substance?

How do you calculate the concentration of a solution?

Relative mass

Relative mass is in terms of ^{12}C.

Relative atomic mass, A_r, is the weighted average mass of an element compared to $\frac{1}{12}$ the mass of an atom of carbon-12. A_r has no units.

Relative molecular mass (for covalent substances) and relative formula mass (for ionic compounds), M_r, is the sum of all the A_r for each atom in the formula. M_r has no units.

> **Calculate the relative formula mass of calcium nitrate, $Ca(NO_3)_2$.**
> Relative formula mass
> $= 40.1 + 2 \times (14.0 + 3 \times 16.0)$
> $= 164.1$

> Remember that everything inside a set of brackets is multiplied by the number outside the brackets, so $Ca(NO_3)_2$ contains 1 calcium, 2 nitrogen and 6 oxygen atoms.

The mole (n)

In one mole of any substance there are exactly 6.022×10^{23} particles (Avogadro's number). Mole has the unit mol.

The molar mass of a substance is the mass of one mole. It is the same number as the A_r or M_r of the substance but measured in grams.

12 g of carbon 6.022×10^{23} carbon atoms

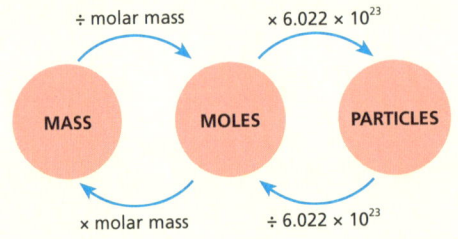

$\div$ molar mass $\times 6.022 \times 10^{23}$

MASS   MOLES   PARTICLES

$\times$ molar mass $\div 6.022 \times 10^{23}$

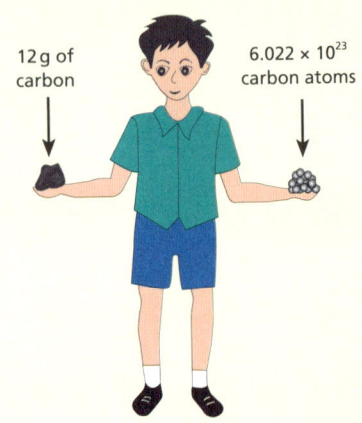

Formulae

Amount of substance (mol) $= \dfrac{\text{mass (g)}}{\text{relative atomic mass}}$ $n = \dfrac{m}{A_r}$

Amount of substance (mol) $= \dfrac{\text{mass (g)}}{\text{relative molecular mass}}$ $n = \dfrac{m}{M_r}$

Amount of substance (mol) $= \dfrac{\text{mass (g)}}{\text{relative formula mass}}$

Amount of substance (mol) $=$ concentration (mol dm^{-3}) $\times$ volume (dm^3) $n = CV$

where $1000\,\text{cm}^3 = 1$ litre $= 1\,\text{dm}^3$

> **Calculate the mass of sodium chloride in 10.00 cm^3 of 2.00 mol dm^{-3} solution.**
> $n = CV = 2.00 \times \frac{10.00}{1000} = 0.02$ moles
> M_r of NaCl $= 23.0 + 35.5 = 58.5$
> $m = M_r \times n = 0.02 \times 58.5 = 1.17$ g

Significant figures (s.f.)

Calculated results can only be reported to the limits of the least accurate measurement.

	1st s.f.	2nd s.f.	3rd s.f.	4th s.f.
4620	4	6	2	0
0.0058	5	8	No further s.f.	

Summary

The mole

(1) When driving in the UK, the legal limit for ethanol ($M_r = 46.0$) is 80 mg per 100 cm^3 of blood.

Calculate the concentration of ethanol in blood in mol dm^{-3}. Tick **one** box. [1]

1.74 ☐ 1.74×10^{-2} ☐ 1.73×10^2 ☐   0.173 ☐

(2) How many oxygen atoms are present in 32.0 g of oxygen gas? Tick **one** box.

The Avogadro constant, $L = 6.022 \times 10^{23}$ mol^{-1} [1]

6.022×10^{23} mol^{-1} ☐ 12.04×10^{23} mol^{-1} ☐ 1.204×10^{24} mol^{-1} ☐ 1.204×10^{-24} mol^{-1} ☐

(3) Hydrogen peroxide, H_2O_2, is sold commercially at a concentration of 1.76 mol dm^{-3}.

When used in hairdressers, it is diluted to 0.050 mol dm^{-3}.

a) Calculate the M_r of hydrogen peroxide. Give your answer to the appropriate precision. [1]

....

b) Calculate the number of moles of hydrogen peroxide in 25.00 cm^3 of commercially-sold hydrogen peroxide. [3]

....

c) Calculate the mass of hydrogen peroxide in 20.00 cm^3 of diluted hydrogen peroxide when used by a hairdresser. Give your answer to 2 significant figures. [2]

....

(4) 22.70 cm^3 of 0.01 mol dm^{-3} sodium carbonate solution, Na_2CO_3, were completely neutralised by 25.00 cm^3 of hydrochloric acid, HCl, to make a salt, water and carbon dioxide gas.

a) Write a balanced symbol equation for this reaction. [2]

....

b) Calculate the number of moles of sodium carbonate solution that were used. [2]

....

c) The acid and base react at a ratio of 2 : 1.

Calculate the concentration of the hydrochloric acid. Give your answer in standard form to 2 significant figures. [3]

Ideal gas, and molecular and empirical formulae

Key words

Key questions

What is an ideal gas?

What is the ideal gas equation?

What is molecular formula used for?

What is the formula type used to describe ionic substances?

Ideal gas

The properties of gases are difficult to predict so they are modelled using an ideal gas, where all gas particles:

- are identical with no volume
- rarely touch and move in all directions with a range of speeds
- have no interactions with each other
- collide elastically.

The ideal gas equation:

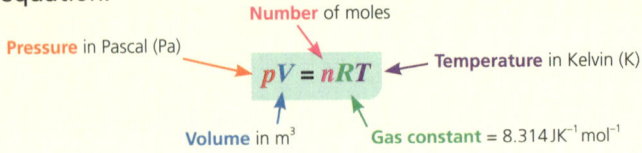

Number of moles

Pressure in Pascal (Pa)

Temperature in Kelvin (K)

$$pV = nRT$$

Volume in m³

Gas constant = 8.314 JK⁻¹ mol⁻¹

You will always be given the value of R in the exam.

To convert:

- pressure in kilopascals (kPa) to pascals (Pa), multiply by 1000
- temperature in degrees Celsius (°C) to Kelvin (K), add 237.

How many moles of gas are present in 200 cm³ of gas at 70°C and 101 kPa?

Rearrange the equation to make the required answer the subject of the equation.	Convert all information into the standard units.	Substitute values into the equation.
$pV = nRT$ $n = \frac{pV}{RT}$	$200\,cm^3 = \frac{200}{1\,000\,000} = 2 \times 10^{-4}\,m^3$ $70°C = 70 + 273 = 343\,K$ $101\,kPa = 101 \times 1000 = 101\,000\,Pa$	$n = \frac{101\,000 \times 0.0002}{8.31 \times 343}$ $= 0.0071$ moles of gas

---

Molecular and empirical formulae

The **molecular formula** of a covalent substance is the actual number of atoms of each element in a molecule.

The **empirical formula** is the smallest whole number ratio of atoms in a compound.

The molecular formula can be derived from the empirical formula by using relative mass.

All ionic compound formulae are reported as empirical formulae.

What is the molecular formula of a molecule with empirical formula CH_2 and M_r 84?

Mass of empirical formula

$12 + 2 = 14$

Divide molecular mass by mass of empirical formula

$\frac{84}{14} = 6$

Multiply the empirical formula by this factor

$6 \times CH_2 = C_6H_{12}$

Determine the empirical formula for a compound made from 36.45 g of magnesium, 48.15 g of sulfur and 96.00 g of oxygen.

Element	Mg	S	O
mass (g)	36.45	48.15	96.00
mole ratio (divide by atomic mass)	$\frac{36.45}{24.3} = 1.5$	$\frac{36.45}{24.3} = 1.5$	$\frac{36.45}{24.3} = 1.5$
element ratio (divide largest by smallest)	$\frac{1.5}{1.5} = 1$	$\frac{1.5}{1.5} = 1$	$\frac{6}{1.5} = 4$
empirical formula	$MgSO_4$		

Summary

Ideal gas, and molecular and empirical formulae

1. What is the volume occupied by 1.43 g of the gas methane, CH_4, at 100 kPa and 273 K? Tick **one** box. [1]

   2.03 dm³ ☐

   2.10 dm³ ☐

   2.03 cm³ ☐

   2.10 cm³ ☐

2. Which compound has the empirical formula CH_2? Tick **one** box. [1]

   Methane ☐   Propene ☐   Propane ☐   Butane ☐

3. A hydrocarbon was analysed and found to contain 83.7% carbon.

   a) Determine the empirical formula of the hydrocarbon. [3]

   b) The relative formula mass, M_r, for the hydrocarbon is 86.0

   Determine the molecular formula of the hydrocarbon. [3]

4. Sulfur can form more than one stable oxide compound.

   0.201 g of one sulfur oxide was analysed at 415 K and found to occupy a volume of 127 cm³ at a pressure of 103 kPa.

   State the ideal gas equation and use it to calculate the number of moles of A in the sample. Hence, calculate the relative molecular mass of A. The gas constant $R = 8.31 \, JK^{-1} mol^{-1}$ [5]

Chemical equations

Key words

Key questions

Why must a full equation be balanced?

What is the function of an ionic equation?

Why is it important to have high atom economy?

Why is it important to have high percentage yield?

Full equations

Full equations (balanced symbol equations):
- show the formula of all the substances and their stoichiometric coefficient (for balancing)
- can be used to determine the mass, moles or volume of one substance if information about the other substances are given.

> $CaCO_3(s) \longrightarrow CaO(s) + CO_2(g)$
>
> **If 250 kg of $CaCO_3$ was heated to produce CaO, what volume of CO_2 (g) would be released?**
>
> To convert from kg to g, multiply by 10^3
>
> M_r $CaCO_3 = 100$ moles in 250 kg $= \frac{250}{100} = 2.5 \times 10^3$
>
> Mole ratio $CaCO_3 : CO_2 = 1:1$ moles $CO_2 = 2.5 \times 10^3$
>
> To convert from moles to volume, multiply by 24
>
> Volume $CO_2 = 2.5 \times 10^3 \times 24 = 60\,000\,dm^3$

> **Air at standard temperature and pressure (STP) is 21% oxygen by volume. How many moles of oxygen are present in a room with a volume of $72\,000\,dm^3$?**
>
> 21% of $72\,000\,dm^3 = \frac{72\,000}{100} \times 21 = 15\,120\,dm^3$ To convert from volume gas at rtp to moles, divide by 24
>
> $\frac{15\,120}{24} = 630$ moles of oxygen

Ionic equations

Ionic equations show only the ions or complete covalent or insoluble ionic compounds involved in the chemical change. For a neutralisation reaction where an acid is neutralised by an alkali, the ionic equation is always:

$$H^+(aq) + OH^-(aq) \longrightarrow H_2O(l)$$

Atom economy

Atom economy is a way of measuring how much of the starting material makes it into the useful product.

$$\text{Atom economy} = \frac{\text{sum of } M_r \text{ of desired products}}{\text{sum of } M_r \text{ of reactants}} \times 100\%$$

> **Manufacture of chloroethane from ethanol**
>
> $C_2H_6O + HCl \longrightarrow C_2H_5Cl + H_2O$
>
> M_r of desired product $C_2H_5Cl = 64.5$
>
> M_r $C_2H_6O = 46$ M_r $HCl = 36.5$
>
> Atom economy $= \frac{64.5}{82.5} \times 100\% = 78.2\%$

Industry wants high atom economy because it is:
- more sustainable (there is less waste and resources are conserved)
- often a cheaper process (uses less energy and resources).

Percentage yield is a measure of how much of the desired product is collected compared to the maximum mass that could have been made:

$$\% \text{ yield} = \frac{\text{actual yield}}{\text{theoretical yield}} \times 100\%$$

> **What is the percentage yield of ammonium sulfate if the starting mass of ammonia is 100 g and the final mass of ammonium sulfate is 364 g?**
>
> $2NH_3(aq) + H_2SO_4(aq) \longrightarrow (NH_4)_2SO_4(aq)$
>
> Calculate the maximum mass from the equation.
>
> M_r $NH_3 = 17$
>
> $\frac{100}{17} = 5.88$ starting moles of ammonia
>
> Mole ratio $= 2:1$
>
> Moles of $(NH_4)_2SO_4 = 0.5 \times 5.88 = 2.94$
>
> M_r of $(NH_4)_2SO_4 = 132.1$
>
> Max. mass $(NH_4)_2SO_4 = 132.1 \times 2.94 = 388\,g$
>
> Divide the actual mass of product by the maximum yield and multiply by 100%
>
> $\frac{364}{388} \times 100\% = 93.7\%$

Summary

Chemical equations

1. Which reaction has the highest atom economy for making ethanol? Tick **one** box. [1]

Hydration of ethene ☐

Fermentation of glucose ☐

Hydrolysis of ethyl methanoate ☐

Hydrolysis of ethyl ethanoate ☐

2. 16.0 g of methane (M_r = 16) reacts with chlorine to make chloromethane (M_r = 50.5). The percentage yield is 65.0%

What mass, in grams, of chloromethane is produced? Tick **one** box. [1]

32.83 g ☐

32.82 g ☐

50.50 g ☐

65.65 g ☐

3. CFCs like trichlorofluoromethane (CCl_3F) can be used as refrigerant.

The manufacture of new CCl_3F is restricted.

The equation for the reaction is: $SbF_3Br_2 + CCl_4 \longrightarrow CCl_3F + SbF_2Br_2Cl$

Calculate the percentage atom economy for the production of CCl_3F in this reaction. Give your answer to 2 significant figures. [3]

4. A scientist analysed an unknown substance, X, using test tube reactions.

a) The scientist added a solution of acidified silver nitrate solution to substance X and observed a white precipitate of silver chloride.

Write an ionic equation for this reaction. [1]

b) The scientist added dilute sodium hydroxide solution to substance X and observed a white precipitate of magnesium hydroxide.

Write an ionic equation for this reaction. [1]

c) The scientist added a small piece of calcium metal to a solution of substance X.

Write a balanced symbol equation for this reaction. Include state symbols. [2]

Ionic bonding

Key words

Key questions

What is an ionic bond?

How do atoms become ions?

What is an ionic lattice?

What is a compound ion?

Ionic bonds

Ionic bonds are the electrostatic force of attraction between oppositely charged ions. **Dot and cross** diagrams can be used to show the ions in a compound:

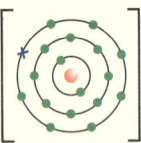
Cl^- ion $[2,8,8]^-$

Mg^{2+} ion $[2,8]^{2+}$

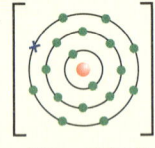
Cl^- ion $[2,8,8]^-$

To work out the ionic compound formula
1. Write the formula of each ion next to each other, starting with the metal: Mg^{2+} Cl^{1-}
2. You need to circle the numbers and have arrows that cross down:
   Mg^{2+} Cl^{1-} becomes Mg_1Cl_2
3. Remove any 1s and cancel down: $MgCl_2$

$Cl^- = 1s^2\ 2s^2\ 2p^6\ 3s^2\ 3p^6$ **isoelectric** with Ar

$Mg^{2+} = 1s^2\ 2s^2\ 2p^6$ isoelectric with Ne

Atoms form ions by losing or gaining electrons to have a complete outer shell:

- Metal atoms form **cations** by oxidation, e.g. $Mg \longrightarrow Mg^{2+} + 2e^-$
- Non-metal atoms form **anions** by reduction, e.g. $Cl_2 + 2e^- \longrightarrow 2Cl^-$

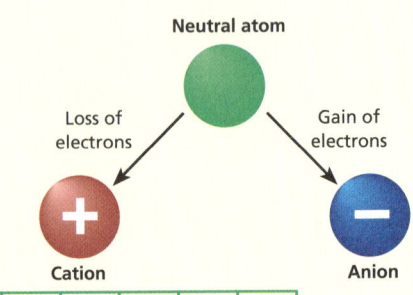

Neutral atom

Loss of electrons — Cation

Gain of electrons — Anion

Group	1	2	3	4	5	6	7	0
Charge on ion	1+	2+	3+	*	3–	2–	1–	*

* Does not form ions

Ionic compounds form giant structures called **lattices**.

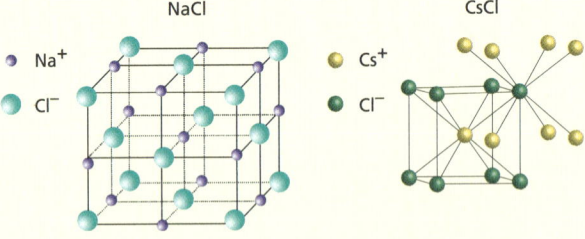

NaCl

- Na$^+$
- Cl$^-$

CsCl

- Cs$^+$
- Cl$^-$

Compound ions are small groups of atoms bonded together with a charge.

NO_3^- nitrate(V)	
HCO_3^- hydrogen carbonate	
CO_3^{2-} carbonate	
SO_4^{2-} sulfate	
NH_4^+ ammonium	
OH^- hydroxide	

NaOH (s)

- Sodium, Na
- Oxygen, O
- Hydrogen, H

Summary

Ionic bonding

1. Element Z forms a hydroxide with formula ZOH.

   Which of these could represent the electronic configuration of an atom of Z? Tick **one** box. [1]

   $[Ne]3s^1$ ☐

   $[Ne]3s^2$ ☐

   $[Ne]3s^23p^1$ ☐

   $[Ne]3s^13p^2$ ☐

2. What is the formula of magnesium nitrate(V)? Tick **one** box. [1]

   $MgNO_3$ ☐

   Mg_2NO_3 ☐

   $Mg(NO_3)_2$ ☐

   $MgNO_2$ ☐

3. Silver reacts with fluorine to form silver fluoride (AgF).

   Silver fluoride has a structure similar to that of sodium chloride.

   a) Describe the bonding involved in silver fluoride. [2]

   b) Name the structure that silver fluoride forms. [1]

   c) Silver can also form a compound with sulfate ions.

   Deduce the formula of silver sulfate. [1]

4. Lithium bromide forms a crystalline solid at room temperature.

   a) On the diagram, mark the charge on each ion. [1]

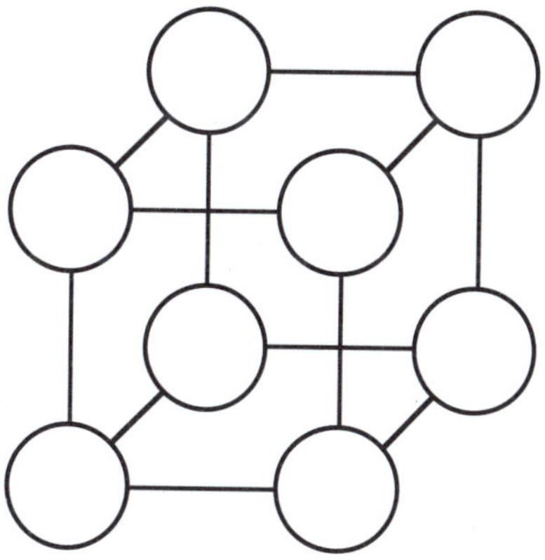

   b) Give the electronic configuration of the lithium ion. [1]

   c) Deduce the noble gas which is isoelectric with the bromide ion. [1]

Covalent bonding

Key words

Key questions

What is a covalent bond?

Covalent bonds

A single **covalent bond** is formed by a shared pair of electrons between the nuclei of two atoms. Each covalent bond is represented by a dot and cross in a dot-and-cross diagram or by a line in a displayed formula.

	Chlorine Cl$_2$	Oxygen O$_2$	Nitrogen N$_2$
Dot and cross	Cl Cl	O O	N N
Displayed formula	Cl – Cl	O = O	N ≡ N

The maximum number of electrons that an element can have available for bonding depends on its Period number:

- Period 1 can form a maximum of one covalent bond.
- Period 2 can form a maximum of four covalent bonds.
- Period 3 can have more than four covalent bonds.

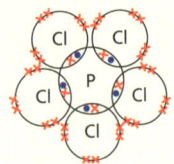

PCl$_5$: all 5 electrons from the outer bonding shell of phosphorus are used in covalent bonds.

What is a dative (co-ordinate) bond?

A **dative (co-ordinate) bond** is a shared pair of electrons where both electrons are donated by one atom.

This is represented as an arrow, where the arrow points away from the atom that is donating the pair of electrons.

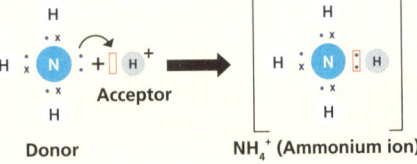

NH$_4^+$ (Ammonium ion)

How do you represent a dative (co-ordinate) bond in a displayed formula?

Covalent substances can form two structures:

- Macromolecular, e.g. diamond

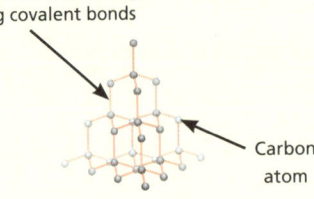

Each carbon atom is held by four strong covalent bonds

Carbon atom

- Simple molecular, e.g. chlorine

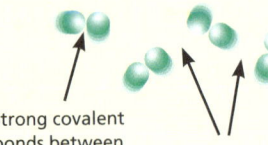

Strong covalent bonds between the atoms in a molecule

Weak intermolecular forces of attraction between molecules. The forces are up to 10% of the strength of a chemical bond

✔ Summary

Covalent bonding

1 The equation for a reaction is $H_2O + H^+ \longrightarrow H_3O^+$

What type of interaction forms in this reaction? Tick **one** box. [1]

Ionic bond ☐

Covalent bond ☐

Co-ordinate bond ☐

Hydrogen bond ☐

2 Which molecule is not able to form a co-ordinate bond with another species? Tick **one** box. [1]

NH_3 ☐ H_2O ☐ BH_3 ☐ C_2H_4 ☐

3 Ammonia is a basic gas at room temperature.

a) Name the bonding present in a molecule of ammonia. [1]

....

b) Boron trichloride (BCl_3) reacts with ammonia to make a compound with a co-ordinate bond.

Define the term 'co-ordinate bond'. [2]

....

....

....

c) Ammonia can react with hydrogen chloride gas to make ammonium chloride.

Draw the displayed formula of the ammonium ion. [2]

4 Chlorine forms compounds with many other elements. Chlorine reacts with methane to form a mixture of compounds, including chloromethane, CH_3Cl.

a) State the type of bond between C and Cl in CH_3Cl and describe how this bond is formed. [2]

Type of bond: ....

How bond is formed: ....

....

b) Explain why chloromethane cannot form a co-ordinate bond with another species. [1]

....

....

Metallic bonding

Spec. ref. 3.1.3.3

Key words

Key questions

What is a metallic bond?

Which substances can form metallic bonds?

What affects the strength of metallic bonds?

Metallic bonds

Only **pure metals** and **alloys** can make **metallic bonds**.

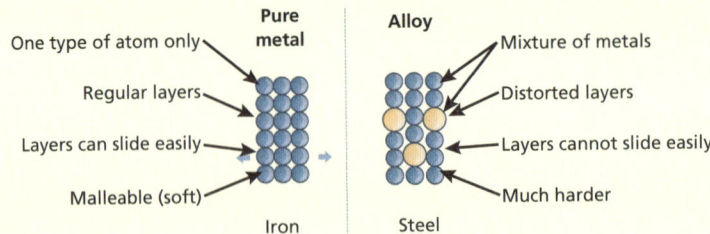

The metal atoms form a **giant structure** (metal crystal) where the outer shell of electrons in each atom merge. This allows the electrons to be **delocalised** and move freely throughout the structure.

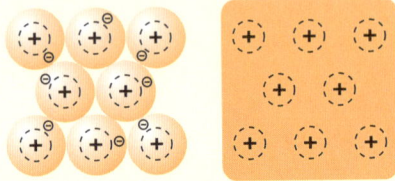

Metallic bonds are the attraction between delocalised electrons and the positive ion in the giant structure.

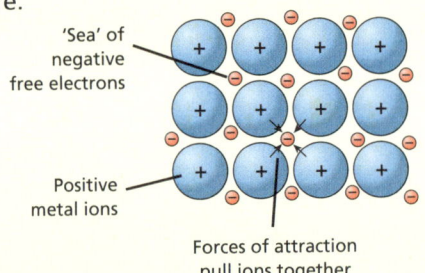

The strongest metallic bonds have:

- a large effective nuclear charge
- lots of outer shell electrons and so a large number of delocalised electrons
- small metal ions.

Aluminium and sodium are both metal elements. Suggest which metal has the stronger metallic bonds.

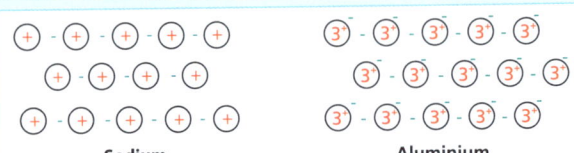

Aluminium has stronger bonds than sodium because aluminium forms smaller ions with a higher charge and releases a greater number of delocalised electrons in the metallic structure. This results in a larger electrostatic force of attraction between the delocalised electrons and the metal ions compared with sodium metal.

Summary

Spec. ref. 3.1.3.3

Metallic bonding

1. Which one of the following substances does not contain delocalised electrons? Tick **one** box. [1]

   Sodium ☐ Steel ☐ Graphite ☐ Diamond ☐

2. Which one of the following substances would likely have the strongest metallic bonds? Tick **one** box. [1]

   Sodium ☐ Magnesium ☐ Aluminium ☐ Gallium ☐

3. Lithium and potassium are elements that are found in Group 1 of the Periodic Table.

   a) Describe the bonding in a sample of lithium metal. [2]

   b) Name the structure formed by potassium atoms in a sample of solid sodium. [1]

   c) Suggest which metal, lithium or potassium has the stronger type of bonding. Give a reason for your answer. [3]

4. Sodium is found in Period 3 of the Periodic Table.

   a) Give the electronic structure of a sodium ion. [1]

   b) Name the type of bonding present in a sample of pure sodium. [1]

   c) Draw a diagram that shows how the particles are arranged in a crystal of sodium. You should identify the particles and show a minimum of six particles in a two-dimensional diagram. [2]

Bonding and physical properties

🔑 Key words

Summary of crystal structures

Crystal structure	Description of bonding	Example	Melting point	Conductivity
Ionic	Electrostatic force of attraction between oppositely charged ions.	Sodium chloride ◯ Cl⁻ • Na⁺	High – many strong bonds to be overcome.	(aq) and (l) when ions are free to move.
Metallic	Electrostatic force of attraction between metal ions and delocalised electrons.	Magnesium e⁻ Mg²⁺ e⁻ Mg²⁺ e⁻ Mg²⁺ e⁻ e⁻ e⁻ e⁻ e⁻ Mg²⁺ Mg²⁺ Mg²⁺ e⁻ e⁻ e⁻	High – many strong bonds to be overcome.	Always as delocalised electrons are free to move and carry the charge.
Macromolecular (giant covalent)	Covalent with a shared pair of electrons between nuclei.	Diamond Graphite	High – many strong bonds to be overcome.	Only if there are delocalised electrons e.g. graphite.
Molecular	Covalent with a shared pair of electrons between nuclei.	Iodine Ice	Low – no bonds are overcome only the forces between the molecules.	Never.

❓ Key questions

What are physical properties?

Why do some substances have a high melting point?

How do substances conduct?

What is latent heat?

Latent heat

Energy or latent heat is transferred into or out of a system for a substance to change state. When a substance melts, the energy needed is called the **enthalpy change**. When a substance boils, the energy change is called the **enthalpy change of vaporisation**.

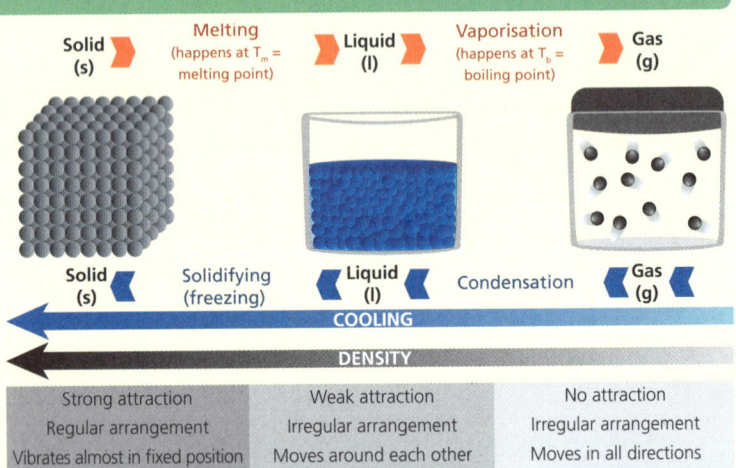

Solid (s) ➤ Melting (happens at Tₘ = melting point) ➤ Liquid (l) ➤ Vaporisation (happens at T_b = boiling point) ➤ Gas (g)

Solid (s) ◀ Solidifying (freezing) ◀ Liquid (l) ◀ Condensation ◀ Gas (g)

COOLING

DENSITY

Strong attraction	Weak attraction	No attraction
Regular arrangement	Irregular arrangement	Irregular arrangement
Vibrates almost in fixed position	Moves around each other	Moves in all directions
It has its own shape	It takes container shape	It has no shape
Rigid	Not rigid	Not rigid
Fixed volume	Fixed volume	Not fixed volume

✔ Summary

Bonding and physical properties

1. Which one of the following substances forms a macromolecular structure? Tick **one** box. [1]

   Silicon dioxide ☐   Magnesium ☐   Sodium chloride ☐   Argon ☐

2. Which structure is formed by water? Tick **one** box. [1]

   Covalent ☐   Macromolecular ☐   Molecular ☐   Lattice ☐

3. Magnesium metal can react in an exothermic reaction with bromine to make magnesium bromide.

   a) Explain, in terms of structure and bonding, why magnesium bromide has a high melting point. [4]

   b) Explain, in terms of structure and bonding, in which state magnesium bromide can conduct electricity. [3]

4. This question is about the structure and bonding of water.

   a) State the structure and bonding found in water. [2]

   Structure: ....

   Bonding: ....

   b) A sample of ice is heated.

   Explain why the temperature does not increase when ice is melting. [3]

   c) Draw a diagram that shows how the particles are arranged in a gaseous sample of water. You should identify the particles and show a minimum of six particles in a two-dimensional diagram. [2]

Shapes of molecules

Key words

Key questions

How do outer shell electron pairs on a central atom interact?

How do you determine the shape of a molecule or ion?

What happens to the bond angles in a molecule when a bonding pair of electrons is replaced with a lone pair of electrons?

Electron pair repulsion and bond angles

Valence shell electron pair repulsion (VSEPR) states that:
- pairs of electrons exist as charge clouds that repel each other
- outer shell electron pairs are as far apart as possible to minimise repulsion
- lone pair – lone pair repulsion > lone pair – bond pair repulsion > bond pair – bond pair repulsion.

Shape	Examples
Linear Bonding pairs: 2 Lone pairs: 0	Cl—Be—Cl 180°
Trigonal planar Bonding pairs: 3 Lone pairs: 0	F–B(F)(F) 120° [O–C(O)(O)]²⁻ 120°
Tetrahedral Bonding pairs: 4 Lone pairs: 0	109.5° H–N(H)(H)H [109.5° H–N(H)(H)H]⁺
Trigonal pyramidal Bonding pairs: 3 Lone pairs: 1	H–N(H)(H) 107°

Shape	Examples
Bent Bonding pairs: 2 Lone pairs: 2	H–O(··)(··)–H 104.5°
Trigonal bipyramidal Bonding pairs: 5 Lone pairs: 0	Cl–P(Cl)(Cl)(Cl)Cl 120° 90°
Octahedral Bonding pairs: 6 Lone pairs: 0	F–S(F)(F)(F)(F)F 90° 90°

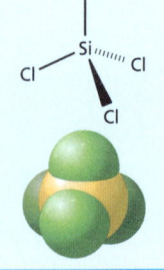

Determine the shape of SiCl₄

1. Identify the central atom: **Si**
2. Determine how many electrons it has in the outer shell (Group number): **Four**
3. Add an electron for every covalent bond on the central atom: **Four**
4. Determine the number of pairs of electrons on the central atom: **Eight electrons in total, so four pairs**
5. Determine the shape: **tetrahedral**

Lone pairs on a central atom reduce the bond angles by about 2.5° per pair, compared to bonding pairs.

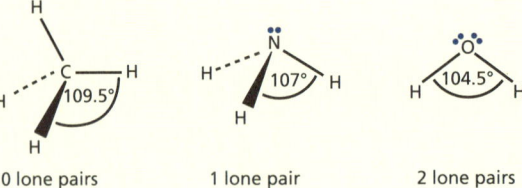

0 lone pairs   1 lone pair   2 lone pairs

✔ Summary

Shapes of molecules

1. Which one of the following substances has a tetrahedral shape? Tick **one** box. [1]

   Ammonium ion ☐ Ammonia ☐ Hydroxonium ion ☐ Water ☐

2. Which one of the following substances has a bond angle of 109.5°? Tick **one** box. [1]

   C(graphite) ☐ C(diamond) ☐ Water ☐ Silver chloride ☐

3. Fluoroantimonic acid ($HSbF_6$) contains two ions, SbF_6^- and H_2F^+.

   a) Determine the shape and bond angles for SbF_6^-. [2]

   ....

   ....

   ....

   b) i) Describe the structure of H_2F^+. [1]

   ....

   ii) Draw a diagram of H_2F^+. Include any lone pairs that influence the shape. [1]

   iii) Suggest the bond angle between the bonding pairs of electrons in H_2F^+. [1]

   ....

4. Explain how the electron pair repulsion theory can be used to deduce the shape of, and the bond angle in, a hydroxonium ion (H_3O^+). [6]

   ....

   ....

   ....

   ....

   ....

   ....

   ....

   ....

Bond polarity and intermolecular forces

Key questions

What is electronegativity?

Bond polarity

Different atoms attract electrons by different amounts. This leads to a bonding continuum.

Non-polar covalent
Electrons are shared equally

Polar covalent
Electrons are shared unequally

Ionic
Electrons are transferred

Cl Cl   H Cl   Na⁺ Cl⁻

Increasing ionic character →

		Difference in electronegativity
0.4	1.7	

Electronegativity is a measurement of the ability of an atom to withdraw electron density from a covalent bond.

H 2.2						
Li 1.0	Be 1.6	B 1.8	C 2.5	N 3.0	O 3.4	F 4.0
Na 0.9	Mg 1.3	Al 1.6	Si 1.9	P 2.2	S 2.6	Cl 3.2
K 0.8						Br 3.0
						I 2.7

Only molecules of elements form non-polar bonds because they have an equal sharing of electrons. All covalent compounds will have unsymmetrical electron distribution but if the difference is between 0.3 and 1.7, a polar bond is formed. If the difference in electronegativity is >1.7, the compound is said to be ionic.

Individual bonds can be polar but if they are in a symmetric molecule, the molecule will not be polar overall.

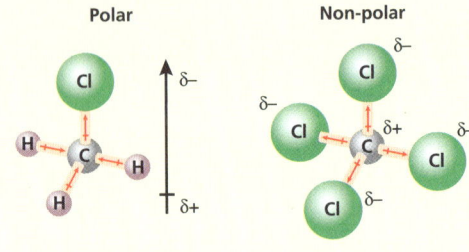

How do polar bonds form?

Intermolecular forces

Which intermolecular force of attraction is always present in molecular substances?

There are covalent bonds between atoms in a simple molecular substance and forces of attraction between the molecules. All molecular substances will have weak (between 0.4 to 4 kJ mol⁻¹) induced dipole–dipole (van der Waals, dispersion, London) forces between the molecules. These forces increase with the number of electrons as the bond is more easily polarisable.

With molecules that have a permanent dipole, there will be permanent dipole–dipole forces between the molecules with a strength of between 5 to 20 kJ mol⁻¹

What is the order of strength for the intermolecular forces?

For molecules that have a O, F or N atom attached to a H atom, hydrogen bonding will be present.

Hydrogen bonding is the strongest intermolecular force with a strength equal to about 10% of a covalent bond. Substances with hydrogen bonds have a much higher boiling point than similar substances with permanent dipole forces.

Induced dipole-dipole attraction between molecules.

As electrons are always moving, sometimes there are more electrons at one end of the molecule

Randomly, there can be fewer electrons at this end of the molecule

$\delta-$ | $\delta+$
Br .... Br

$\delta+$ $\delta-$
Br .... Br

Due to the effect of the electrons in the other molecule, electrons are repelled from this end of the molecule

Due to the effect of the electrons in the other molecule, electrons are attracted to this end of the molecule

Less electronegative   More electronegative

$\delta+$ $\delta-$
H — Cl

$\delta-$ Cl — H $\delta+$

Dipole-dipole attraction between molecules

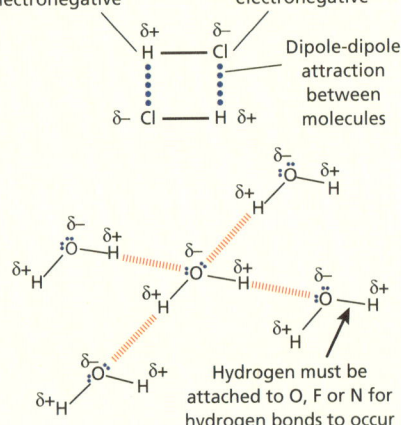

Hydrogen must be attached to O, F or N for hydrogen bonds to occur

Summary

Bond polarity and intermolecular forces

1. Which molecule has a permanent dipole? Tick **one** box. [1]

   NaCl ☐   CCl$_4$ ☐   CH$_3$Br ☐   CBr$_4$ ☐

2. Which bond has the most unsymmetrical electron distribution? Tick **one** box. [1]

   H–F ☐   O–H ☐   H–Cl ☐   H–S ☐

3. Water, H$_2$O, contains polar bonds.

   a) Define the term 'polar bond'. [2]

   ....

   ....

   b) Explain why water forms a polar molecule. [3]

   ....

   ....

   ....

   c) Ammonia, NH$_3$, can dissolve in water by forming hydrogen bonds.

   Draw a diagram to show the hydrogen bonds between water and ammonia molecules. [2]

4. This question is about bromoethane.

   a) Define the term 'electronegativity'. [1]

   ....

   b) Explain the polarity of the C–Br bond in bromoethane. [2]

   ....

   ....

   ....

Enthalpy change and calorimetry

Key words

Key questions

What is enthalpy change?

What are the standard conditions?

How do you calculate enthalpy change from a calorimetry experiment?

What is the difference between exothermic and endothermic reactions?

Measuring enthalpy change

Enthalpy (H) is energy and cannot be measured directly.

Enthalpy change, ΔH, is the heat energy change and can be measured under constant pressure:

Coffee cup calorimetry

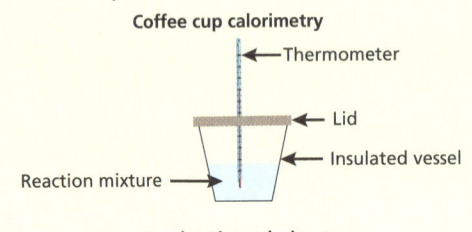

- Thermometer
- Lid
- Insulated vessel
- Reaction mixture

Combustion calorimetry

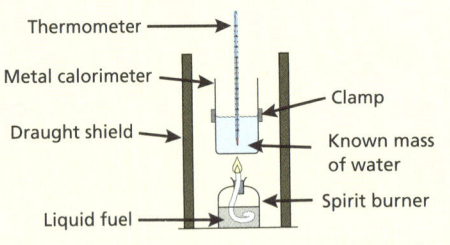

- Thermometer
- Metal calorimeter
- Draught shield
- Clamp
- Known mass of water
- Spirit burner
- Liquid fuel

Measured enthalpy changes are always much lower than calculated because of heat lost to the surroundings.

A student carried out an experiment to measure the enthalpy change of combustion of propanol and obtained the following results:

	Starting	Final
mass of fuel + burner (g)	0.64	0.36
temperature of water (°C)	20.0	26.2
mass of water (g)	200	

Mass of fuel burned = 0.64 − 0.36 = 0.28 g

Temperature change = 26.2 − 20.0 = 6.2°C

Note: 1°C = 1 K So, temperature change = 6.2 K

Specific heat capacity of water = 4.18 J g^{-1} K^{-1}

$q = m \times c \times \Delta T$

Energy change = 200 × 4.18 × 6.2 = 5183 J

$= 5.183\,kJ$

M_r propanol $C_3H_8O = 60$

Moles of propanol burned $= \frac{0.28}{60} = 0.0047$

Energy change per mole propanol $= \frac{5.183}{0.0047}$

$= 1103\,kJ$

Enthalpy change of combustion $= -1103\,kJ\,mol^{-1}$

The accurate value for the standard enthalpy of combustion of propanol is $-2021\,kJ\,mol^{-1}$.

Standard enthalpy changes

Standard enthalpy change, $\Delta H^{\varnothing}$, happens under 100 kPa and 298 K.

Standard molar enthalpy of combustion, $\Delta_c H^{\varnothing}$, is the energy change when one mole of a substance is completely combusted under standard conditions where all substances are in their standard state.

Standard molar enthalpy of formation, $\Delta_f H^{\varnothing}$, is the energy change when one mole of a substance is made from elements in their standard states.

Exothermic reactions have a negative enthalpy change as energy is released from the system to the surroundings.

Endothermic reactions have a positive enthalpy change as energy is taken in by the system from the surroundings.

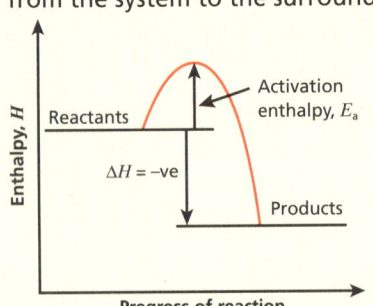

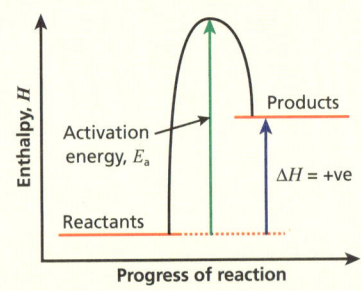

Summary

Enthalpy change and calorimetry

1 Which reaction has an enthalpy change equal to the standard enthalpy of formation of sodium oxide? Tick **one** box. [1]

$4Na(s) + O_2(g) \longrightarrow 2Na_2O(s)$ ☐

$2Na(s) + \frac{1}{2}O_2(g) \longrightarrow Na_2O(s)$ ☐

$2Na(s) + O(g) \longrightarrow Na_2O(s)$ ☐

$2Na^+(s) + O^{2-}(g) \longrightarrow Na_2O(s)$ ☐

2 A student is measuring the enthalpy of combustion of an alcohol.

Which of the following measurements is needed? Tick **one** box. [1]

Starting temperature of the fuel ☐

Temperature of the room ☐

Final temperature of the water ☐

Temperature of the copper calorimeter ☐

3 a) State the standard conditions of enthalpy change. [2]

_____

_____

b) Define the term 'standard enthalpy of combustion'. [3]

_____

_____

_____

c) Write an equation, including state symbols, to show the reaction that takes place when the standard enthalpy of formation for ethanol is measured. [3]

_____

_____

_____

4 A student used calorimetry to investigate the energy in an alcohol. The student completely combusted a 0.0300 mol sample of the alcohol below a copper calorimeter with 0.150 kg of water inside.

The student recorded that the temperature of the water in the copper calorimeter increased from 18.9°C to 78.1°C.

Calculate the enthalpy of combustion, in $kJ\,mol^{-1}$, for the alcohol. The specific heat capacity of water, $c = 4.18\,J\,g^{-1}\,K^{-1}$. [4]

Hess's law

Key words

Key questions

What is Hess's law?

What can Hess's law be used for?

What does $\Delta_f H^\ominus$ mean?

Hess's law

Some **enthalpy changes** cannot be measured directly because a reaction is:

- slow
- difficult to carry out
- complex.

Hess's law states that the enthalpy change of a reaction is independent of the route that is taken and can be used to calculate an enthalpy change.

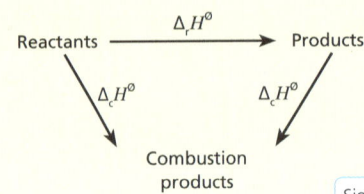

Sigma, Σ, means sum of.

$\Delta_r H^\ominus = \Sigma \Delta_c H^\ominus \text{ (reactants)} - \Sigma \Delta_c H^\ominus \text{ (products)}$

$\Delta_r H^\ominus = \Sigma \Delta_f H^\ominus \text{ (products)} - \Sigma \Delta_f H^\ominus \text{ (reactants)}$

<table>
<tr><td colspan="2">Find the enthalpy of reaction for
C(s) + 2H₂(g) ⟶ CH₄(g)</td></tr>
</table>

Find the enthalpy of reaction for
$$C(s) + 2H_2(g) \longrightarrow CH_4(g)$$

Substance	$\Delta_c H^\circ$ (kJ mol^{-1})
C(s)	−394
H$_2$(g)	−286
CH$_4$(g)	−890

Enthalpy cycle:

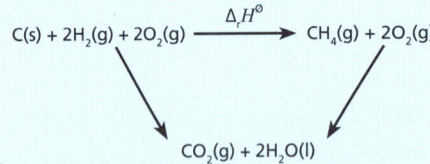

Sum of $\Delta_c H^\ominus$ reactants $= (-394) + 2(-286)$
$= -966\,\text{kJ mol}^{-1}$

Sum of $\Delta_c H^\ominus$ products $= -890\,\text{kJ mol}^{-1}$

$\Delta_r H^\ominus = (-966) - (-890) = -76\,\text{kJ mol}^{-1}$

Note: Enthalpies of combustion are negative so the sum is always a negative value minus a negative.

Find the enthalpy of reaction for
$$CH_4(g) + 2O_2(g) \longrightarrow CO_2(g) + 2H_2O(l)$$

Substance	$\Delta_f H^\circ$ (kJ mol^{-1})
CH$_4$(g)	−76
CO$_2$(g)	−394
H$_2$O(l)	−286

Enthalpy cycle:

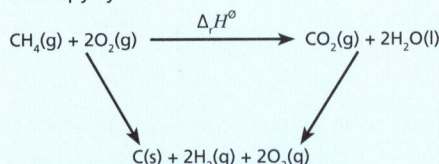

Sum of $\Delta_f H^\ominus$ reactants $= -76\,\text{kJ mol}^{-1}$

Sum of $\Delta_f H^\ominus$ products $= (-394) + 2(-286)$
$= -966\,\text{kJ mol}^{-1}$

$\Delta_r H^\ominus = -(-76) + (-966) = -890\,\text{kJ mol}^{-1}$

Summary

Hess's law

(1) Use the information below to answer the question.

$C(s) + O_2(g) \longrightarrow CO_2(g)$ $\Delta H^{\emptyset} = -393.5\,kJ\,mol^{-1}$

$H_2(g) + \frac{1}{2}O_2(g) \longrightarrow H_2O(g)$ $\Delta H^{\emptyset} = -285.8\,kJ\,mol^{-1}$

$C(s) + 2H_2(g) \longrightarrow C_2H_4(g)$ $\Delta H^{\emptyset} = -84.68\,kJ\,mol^{-1}$

Calculate the value in $kJ\,mol^{-1}$ for the enthalpy of combustion of ethene. Tick **one** box. [1]

1729.08 kJ mol^{-1} ☐   1559.72 kJ mol^{-1} ☐

−1729.08 kJ mol^{-1} ☐   −1559.72 kJ mol^{-1} ☐

(2) Use the information below to answer the question.

$C(s) + O_2(g) \longrightarrow CO_2(g)$ $\Delta H^{\emptyset} = -393.5\,kJ\,mol^{-1}$

$C(s) + \frac{1}{2}O_2(g) \longrightarrow CO(g)$ $\Delta H^{\emptyset} = -110.5\,kJ\,mol^{-1}$

Calculate the value in $kJ\,mol^{-1}$ for the oxidation of carbon monoxide to produce carbon dioxide. Tick **one** box. [1]

−283 kJ mol^{-1} ☐   +504 kJ mol^{-1} ☐

+283 kJ mol^{-1} ☐   −504 kJ mol^{-1} ☐

(3) This question is about enthalpy changes.

a) What does Hess's law state? [1]

_____

_____

b) Iron can react with oxygen to form two different oxides.

The thermodynamic equations for these two reactions of iron with oxygen are shown.

$Fe(s) + \frac{1}{2}O_2(g) \longrightarrow FeO(s)$ $\Delta H^{\emptyset} = -272\,kJ\,mol^{-1}$

$2Fe(s) + \frac{3}{2}O_2(g) \longrightarrow Fe_2O_3(s)$ $\Delta H^{\emptyset} = -822\,kJ\,mol^{-1}$

Use this information to calculate the enthalpy change for this reaction.

$2FeO(s) + \frac{1}{2}O_2(g) \longrightarrow Fe_2O_3(s)$ [3]

(4) The table gives some values of the enthalpy of combustion.

Substance	$C(s)$	$H_2(s)$	$C_4H_8(g)$
$\Delta_c H^{\emptyset}$ (kJ mol^{-1})	−394	−286	−2717

Use this information to calculate the enthalpy of formation for bute-1-ene. [2]

Bond enthalpies

Key words

Key questions

What is mean bond energy?

Why is bond energy always a positive value?

How can you use bond energy to estimate the enthalpy change of a chemical reaction?

Bond enthalpies and calculations

Bond enthalpy is the energy to break one mole of a covalent bond with all species in the gaseous state, e.g. $Cl_2(g) \longrightarrow 2Cl(g) + 243 \, kJ \, mol^{-1}$

Bond enthalpy is:

* an endothermic process
* has a positive enthalpy value.

The same bond will have a different bond enthalpy in different molecules, therefore mean bond energy is used in calculations.

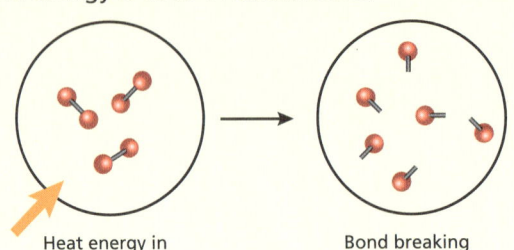

Heat energy in   Bond breaking

Find the enthalpy of reaction for $CH_4(g) + 2O_2(g) \longrightarrow CO_2(g) + 2H_2O(g)$

Sum of bond enthalpies of reactants
$= 4(413) + 2(497) = +2646 \, kJ \, mol^{-1}$

Sum of negative of bond enthalpies of products
$= 2(-805) + 4(-463) = -3462 \, kJ \, mol^{-1}$

Enthalpy of reaction $= +2646 + (-3462)$
$= -816 \, kJ \, mol^{-1}$

Bond	Mean bond enthalpy (kJ mol^{-1})
C–H	413
O=O	497
C=O	805
O–H	463

The calculated enthalpy change of a reaction, using bond energies, will be an estimate because:

* average bond energies are used
* the bond may be in a molecule that is not in the gaseous state.

$\Delta H = \Sigma$ bond energies of the reactants $- \Sigma$ bond energies of the products

For an exothermic reaction:

* ΔH is negative
* Σ bond energies of the reactants $< \Sigma$ bond energies of the products

For an endothermic reaction:

* ΔH is positive
* Σ bond energies of the reactants $> \Sigma$ bond energies of the products

Summary

Bond enthalpies

1 Calculate the value in $kJ\,mol^{-1}$ for the enthalpy of combustion of ethane. Tick **one** box. [1]

	C–H	C–O	H–O
Mean bond disassociation energy ($kJ\,mol^{-1}$)	+412	+360	+463

+1235 $kJ\,mol^{-1}$ ☐ +2059 $kJ\,mol^{-1}$ ☐

−2059 $kJ\,mol^{-1}$ ☐ +12.35 $kJ\,mol^{-1}$ ☐

2 Given the following data:

$Cl_2(g) + H_2(g) \longrightarrow 2HCl(g)$ $\Delta H = -184.6\,kJ\,mol^{-1}$

$H_2(g) \longrightarrow 2H(g)$ $\Delta H = +436\,kJ\,mol^{-1}$

$Cl_2(g) \longrightarrow 2Cl(g)$ $\Delta H = +243\,kJ\,mol^{-1}$

Which one of the following is the bond energy, in $kJ\,mol^{-1}$, for H–Cl? Tick **one** box. [1]

+431.8 ☐ +863.6 ☐

+494.4 ☐ +247.2 ☐

3 Some enthalpy data is given in the table.

Process	Enthalpy change ($kJ\,mol^{-1}$)
$Xe(g) + 2F_2(g) \longrightarrow XeF_4(g)$	−252
$F_2(g) \longrightarrow 2F(g)$	+158

a) Use the data in the table to calculate the bond enthalpy for Xe–F in $XeF_4(g)$. Give appropriate units in your answer. [4]

b) Explain why the calculated value of Xe–F in part a) will be different to published mean bond enthalpy data. [2]

4 The table shows some bond enthalpy data.

Bond	H–H	O=O	H–O
Bond enthalpy ($kJ\,mol^{-1}$)	+436	+496	+464

a) The H–O is a mean bond enthalpy.

State the meaning of the term 'mean bond enthalpy'. [3]

b) Calculate a value for the enthalpy of water formation in the gas phase. [3]

Collision theory and Maxwell–Boltzmann distribution

Key words

Key questions

What is the rate of reaction?

Collision theory

Collision theory is a model used to predict the effect of changing conditions on the rate of a reaction. **Rate of reaction** is a measurement of the quantity of a reactant used or the quantity of product formed over time.

Collision theory states that chemical reactions can only happen if reactant particles collide with enough energy and in the correct orientation. So, most collisions are not successful and do not lead to a reaction.

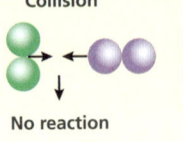

Collision

Correct orientation and sufficient energy so collision occurs

No reaction — Orientation is incorrect

No reaction — Insufficient energy

The rate of reaction can be increased by increasing the concentration of reactant solutions, the pressure of gaseous reactants or the surface area of solid reactants. This is because:
- more particles are available for collision in a given period of time
- although the proportion of successful collisions stays the same, there are more collisions overall
- more successful collisions occur in a given period of time.

Maxwell–Boltzmann distribution

Why do most collisions not lead to a chemical reaction?

The Maxwell–Boltzmann distribution of molecular energies in gases can be used to model the range of energies that particles have at a given temperature.

The area under the curve is the number of molecules present. 60% of molecules in this range

E_a = activation energy of the reaction. The only molecules that can react are those with more energy than E_a

(graph: Number of molecules with a given energy vs Energy)

Increasing the temperature increases the rate of reaction because:
- the particles move faster
- there are more collisions, and more are successful, in a given amount of time
- each collision is likely to be of higher energy.

A small temperature rise, e.g. 10°C, can lead to a large increase in rate of reaction.

Note the change in shape of the distribution curve. But as the number of molecules are the same, the area under both curves is the same

T_1 is a higher temperature than T_2. So, at T_1 there is a greater proportion of molecules with energy greater or equal to E_a and rate of reaction increases

(graph: Number of molecules with a given energy vs Energy, curves T_2 and T_1)

What is the Maxwell–Boltzmann distribution?

Adding a suitable **catalyst** increases the rate of reaction without being changed in chemical composition or amount. A catalyst lowers the **activation energy** and provides an alternative reaction pathway for the reaction.

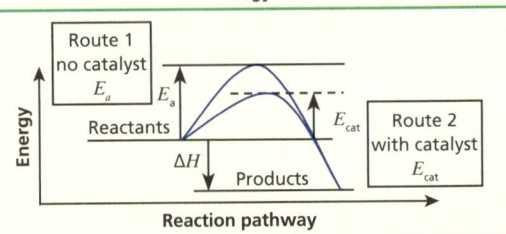

Route 1 no catalyst
E_a
E_a
Reactants
E_{cat}
Route 2 with catalyst
E_{cat}
ΔH
Products

(graph: Energy vs Reaction pathway)

With a catalyst, the proportion of collisions remains the same but the number of collisions that are successful increases.

E_{cat} = activation energy with a catalyst. A greater proportion of molecules now exceeds the lower activation energy ∴ reaction rate increases

(graph: Number of molecules with a given energy vs Energy, E_{cat} and E_a)

Activation energy decreased

Proportion of molecules exceeding the activation energy without a catalyst

Summary

Collision theory and Maxwell–Boltzmann distribution

(1) What does the area under the Maxwell–Boltzmann distribution curve represent? Tick **one** box. [1]

The total number of particles present ☐   The activation energy for a reaction ☐

The total number of reactant particles present ☐   The most probable energy of the particles ☐

(2) Which statement about molecules in a gas is correct? Tick **one** box. [1]

Some gas particles have no energy ☐

All gas particles have the same energy ☐

Average kinetic energy is constant at a fixed temperature ☐

At higher temperature, the mean energy decreases ☐

(3) A student investigated the rate of reaction between hydrochloric acid and pure calcium oxide. The reaction is shown below.

$$2HCl(aq) + CaO(s) \longrightarrow CaCl_2(aq) + H_2O(g) \qquad \Delta H^{\varnothing} = -196.8\,kJ\,mol^{-1}$$

A sketch graph of the student's results is shown.

Use your understanding of collision theory to explain why the student did not obtain a straight line.

[4]

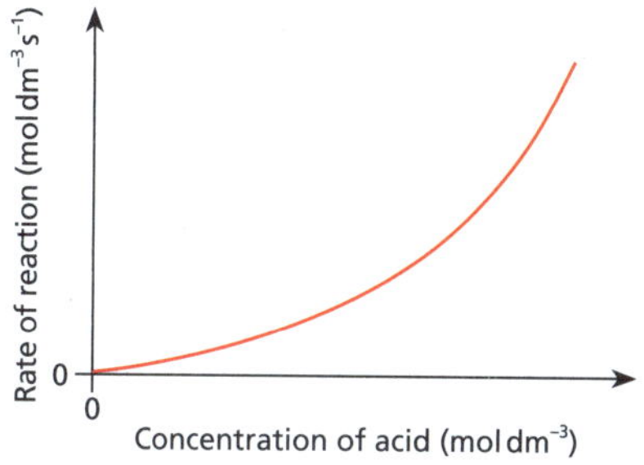

....

....

....

....

(4) The Maxwell–Boltzmann distribution of molecular energies in a sample of gas is shown below.

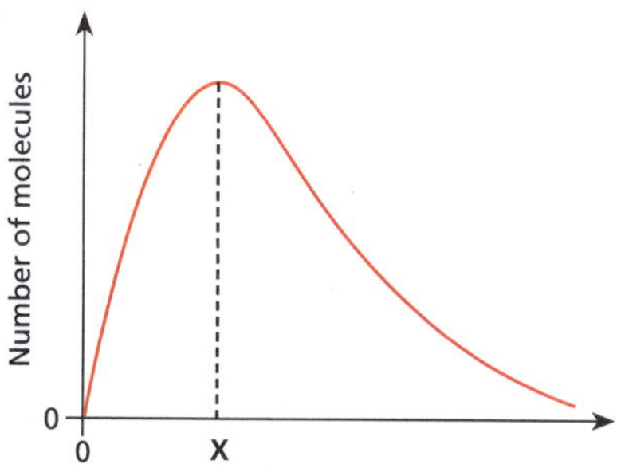

a) Label the x-axis. [1]

b) State what X shows. [1]

....

c) Explain why the curve starts at the origin. [1]

....

d) Explain why the curve never meets the x-axis. [1]

....

e) Draw a Maxwell–Boltzmann distribution curve on the axis above for the same system but 10°C warmer. [2]

Chemical equilibria and Le Chatelier's principle

Key words

Chemical equilibria

A chemical **system** is the reaction that you are studying. Everything else is called the surroundings.

A **reversible reaction** is shown by $\rightleftharpoons$

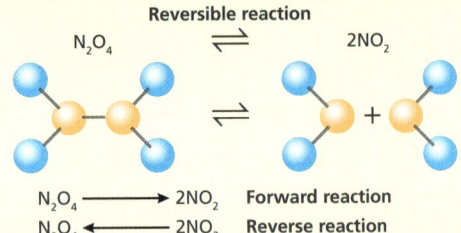

Reversible reaction

$N_2O_4 \rightleftharpoons 2NO_2$

$N_2O_4 \longrightarrow 2NO_2$ Forward reaction
$N_2O_4 \longleftarrow 2NO_2$ Reverse reaction

Key questions

What symbol shows a reversible reaction?

A system made from a reversible reaction can reach **dynamic equilibrium** when:
- the system is closed
- rate of the forward reaction = rate of the reverse reaction
- the concentrations of all substances remain constant.

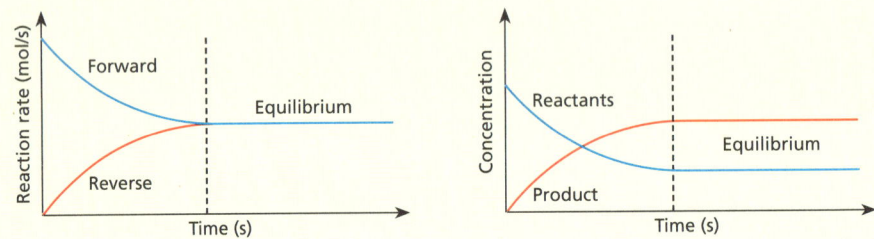

What are the conditions for a reversible reaction to be at dynamic equilibrium?

The addition of a catalyst has no effect on the position of equilibrium. A catalyst makes a system get to dynamic equilibrium quicker as it increases the rate of the forward and reverse reactions by the same amount.

Le Chatelier's principle

Le Chatelier's principle:
- states that the position of equilibrium will shift to oppose any change
- only suggests the effect on yield
- doesn't give any indication on rate of reaction.

Describe the effect on the yield of sulfur trioxide in the Contact Process if the temperature is increased.

$O_2(g) + 2SO_2(g) \rightleftharpoons 2SO_3(g)$ $\Delta H^\varnothing = -1962 \, kJ \, mol^{-1}$

If the temperature is increased, the system will oppose the change and favour the reverse, endothermic reaction. This shifts the position of equilibrium to the left and reduces the yield of sulfur trioxide.

What effect does changing the conditions have on the position of equilibrium for a dynamic equilibrium?

Describe the effect on the yield of ammonia in the Haber Process if the pressure is increased.

$3N_2(g) + 2H_2(g) \rightleftharpoons 2NH_3(g)$ $\Delta H^\varnothing = -92 \, kJ \, mol^{-1}$

If the pressure was increased, the system will oppose the change and favour the products as there are fewer moles of gas. This shifts the position of equilibrium to the right and increases the yield of ammonia.

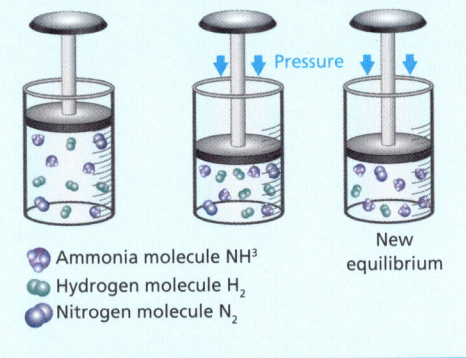

Pressure

New equilibrium

- Ammonia molecule NH^3
- Hydrogen molecule H_2
- Nitrogen molecule N_2

Summary

Chemical equilibria and Le Chatelier's principle

(1) Which of the following conditions has no effect on a system at dynamic equilibrium? Tick **one** box. [1]

Adding a suitable catalyst ☐   Increasing the concentration of the reactants ☐

Removing the products ☐   Decreasing the temperature ☐

(2) The Haber Process can be represented by the equation below.

$$3N_2(g) + 2H_2(g) \rightleftharpoons 2NH_3(g) \qquad \Delta H^\varnothing = -92 \, kJ \, mol^{-1}$$

What is the effect on the yield of ammonia and the rate of reaction in the Haber Process if the temperature is increased? Tick **one** box. [1]

Rate of reaction	Yield of ammonia	Tick one box
Increases	Increases	
Increases	Decreases	
Decreases	Stays the same	
Stays the same	Stays the same	

(3) A homogenous gas phase reaction can be represented as $A + B \rightleftharpoons 3C$

The reaction mixture was allowed to get to dynamic equilibrium at two different temperatures. A sketch graph of the results is shown.

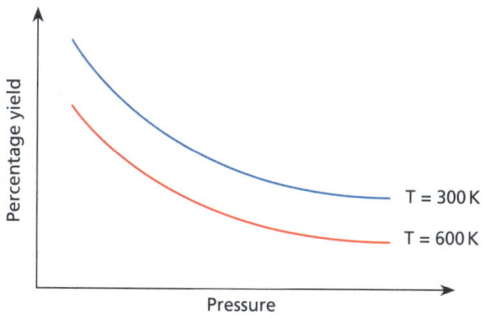

a) Describe and explain the effect on percentage yield as the pressure was increased. [3]

_____

_____

_____

b) Use the graph to determine if the forward reaction was exothermic or endothermic. Explain your answer. [3]

_____

_____

_____

(4) This question is about chemical equilibria.

a) List **two** features of a dynamic equilibrium. [2]

_____

_____

b) Explain the effect of adding a suitable catalyst to a reversible reaction in a closed system. [3]

_____

_____

_____

Equilibrium constant, K_c

🔑 Key words

❓ Key questions

What is an equilibrium constant?

When do equilibrium constants change?

What does '[A]' mean?

What does the value of K_c indicate?

Equilibrium constant, K_c

The equilibrium constant, K_c is a mathematical expression which:
- is deduced from the balanced symbol equation of the reversible reaction
- is not affected by addition of suitable catalysts or changes in concentration or pressure
- is written for homogeneous systems
- changes with temperature as the position of equilibrium will change
- does not indicate the kinetics (rate) of reaction.

$$aA + bB \rightleftharpoons cC + dD$$

$$K_c = \frac{[C]^c\,[D]^d}{[A]^a\,[B]^b}$$

[A], [B], [C] and [D] are concentrations of A, B, C and D respectively

Large values of K_c ($\gg 1$) mean the equilibrium position would be to the right and large yields would be achieved.

Write an expression for the equilibrium constant for the Haber Process.

$$3H_2(g) + N_2(g) \rightleftharpoons 2NH_3(g)$$

Each reactant and product is written as a concentration by using square brackets.	$[NH_3(g)]$, $[H_2(g)]$, $[N_2(g)]$
Each concentration is raised to the power of its coefficient in the reaction equation.	$2NH_3(g) = [NH_3(g)]^2$ $3H_2(g) = [H_2(g)]^3$ $N_2(g) = [N_2(g)]^2$
The products are multiplied together on the top of the equation and the reactants multiplied together on the bottom of the equation.	$\dfrac{[NH_3(g)]^2}{[H_2(g)]^3\,[N_2(g)]^2}$

The concentration of a solid in an equilibrium can be considered to be 1, so is not included in the equilibrium expression.

0.10 moles of $C_2H_5OH(l)$ were mixed with 0.20 moles of $CH_3COOH(l)$. A concentrated acid catalyst was added and the total volume of the mixture was $125\,cm^3$. The reaction was left to come to equilibrium.

$$C_2H_5OH(l) + CH_3COOH(l) \rightleftharpoons C_2H_5OCOCH_3(l) + H_2O(l)$$

The number of moles of CH_3COOH at equilibrium was calculated by titrating against sodium hydroxide and found to be 0.115. Calculate the value of K_c for the reaction at this temperature.

$$K_c = \frac{[C_2H_5OCOCH_3(l)][H_2O(l)]}{[C_2H_5OH(l)][CH_3COOH(l)]}$$ Write the expression

$$\text{units } K_c = \frac{\cancel{mol\,dm^{-3}} \times \cancel{mol\,dm^{-3}}}{\cancel{mol\,dm^{-3}} \times \cancel{mol\,dm^{-3}}}$$ Deduce the units

no units

	C_2H_5OH	CH_3COOH	$C_2H_5OCOCH_3$	H_2O	
Starting amount (mol)	0.1	0.2	0	0	Determine the equilibrium concentrations
Amount at equilibrium (mol)	0.015	0.115	0.085	0.085	
Change in amount (mol)	−0.085	−0.085	+0.085	+0.085	
Concentration ($mol\,dm^{-3}$)	0.015 ÷ 0.125 = 0.12	0.115 ÷ 0.125 = 0.92	0.085 ÷ 0.125 = 0.68	0.085 ÷ 0.125 = 0.68	

$$K_c = \frac{0.68 \times 0.68}{0.12 \times 0.92} = 4.19$$ Complete the calculation

✔ Summary

Equilibrium constant, K_c

1 What are the units for the equilibrium constant in the Haber Process? [1]

$3N_2(g) + 2H_2(g) \rightleftharpoons 2NH_3(g)$

Tick **one** box.

$mol^2\,dm^{-6}$ ☐ $\qquad$ $mol\,dm^3$ ☐ $\qquad$ $mol^{-2}\,dm^6$ ☐ $\qquad$ no units ☐

2 Which statement is correct about the industrial production of ethanol from ethene at 300°C? [1]

$C_2H_4(g) + 2H_2O(g) \rightleftharpoons C_2H_5OH(g) \qquad \Delta H^{\varnothing} = -46\,kJ\,mol^{-1}$

Tick **one** box.

Increasing the pressure decreases the value of K_c ☐

Adding phosphoric catalyst increases the value of K_c ☐

Continuously condensing and removing ethanol increases the value of K_c ☐

Increasing the temperature decreases the value of K_c ☐

3 A homogenous gas phase reaction can be represented as $\qquad A + B \rightleftharpoons 3C$

a) Define the term 'homogeneous'. [1]

b) Write an expression for the equilibrium constant, K_c [1]

c) Deduce the units of the equilibrium constant. [1]

d) Explain the effect on the value of the equilibrium constant by the addition of suitable catalysts. [3]

4 This question is about the equilibrium mixture formed when $ClO(g)$ and $NO_2(g)$ react.

$ClO(g) + NO_2(g) \rightleftharpoons ClONO_2(g)$

5 moles of $ClO(g)$ were mixed with 6 moles of $NO_2(g)$ in a vessel with a volume of $2500\,cm^3$. At equilibrium it was found that there were 2 moles of $NO_2(g)$.

Calculate the equilibrium constant for this reaction. Give units in your answer. [6]

Redox

🔓 Key words

❓ Key questions

Why are oxidation numbers useful?

What happens in a disproportionation reaction?

What are reducing agents?

What are oxidising agents?

Oxidation and reduction

Oxidation state is the formal charge on an atom. It can be calculated using these rules:

- Uncombined elements have an oxidation state of 0.
- The sum of the oxidation numbers of the elements in a compound equals 0.
- The oxidation number of a monatomic ion is the same as its charge.
- The sum of the oxidation numbers of a compound ion equals the overall charge on the ion.
- Fluorine in a compound always has an oxidation state of –1.
- Oxygen has an oxidation state of –2 unless it is in a peroxide (when it is –1).
- Hydrogen has an oxidation state of +1 unless it is in a metal hydroxide (when it is –1).

The oxidation number of sulfur in SO_4^{2-}:
The oxidation number of oxygen is –2.
The sum of the oxidation numbers = charge on the ion = –2
$(4 \times -2) +$ oxidation number of S = –2
Oxidation number of S = +6

The oxidation number of Cr in $K_2Cr_2O_7$:
The oxidation number of a monatomic ion is the charge on the ion: K = +1
The sum of the oxidation numbers in a compound = 0
$(2 \times +1) =$ sum of oxidation numbers in Cr_2O_7 = –2
The sum of the oxidation number for a compound ion = charge on the ion
Charge on Cr_2O_7 = 2–
The oxidation number of oxygen is –2.
$(7 \times -2) + 2 \times$ oxidation number of Cr = –2
$-14 + 2 = 2 \times$ oxidation number of Cr
Oxidation number of Cr = +6

The oxidation number of magnesium and chlorine in $MgCl_2$:
The oxidation number is the same as the charge on the ion.
Magnesium is +2   Chlorine is –1

The oxidation number of nitrogen in NO_2:
The oxidation number of oxygen is –2
The sum of the oxidation number for a compound = 0
Two oxygens = 2×-2 = –4
–4 + oxidation number of nitrogen = 0
Oxidation number of nitrogen = +4

In an **oxidation** reaction, a species could lose electrons and this would increase the oxidation state, e.g. copper(I) ions can be oxidised to copper(II) ions:
$$Cu^+ \longrightarrow Cu^{2+} + e^-$$

In a **reduction** reaction, a species could gain electrons and this would decrease the oxidation state, e.g. iron(III) ions can be reduced to iron(II) ions:
$$Fe^{2+} + e^- \longrightarrow Fe^{2+}$$

Redox reactions

A **redox reaction** is a chemical change in which oxidation and reduction happen at the same time.

In a **disproportion reaction**, one species is oxidised and reduced at the same time.

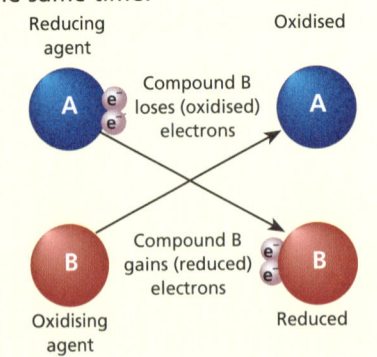

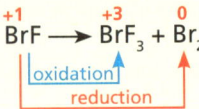

✔ **Summary**

Redox

(1) Which statement correctly describes an oxidising agent? Tick **one** box. [1]

Electron acceptor ☐

Electron donor ☐

Lowers the activation energy ☐

Proton acceptor ☐

(2) Which species is being reduced in the following reaction? Tick **one** box. [1]

$$2Fe_2O_3 + 3C \longrightarrow 4Fe + 3CO_2$$

Fe_2O_3 ☐

C ☐

Fe ☐

CO_2 ☐

(3) Gases in the air can combine in a car engine to form the pollutants NO_x (nitric oxide and nitrogen dioxide). One of the reactions is shown below.

$$2NO + O_2 \longrightarrow 2NO_2$$

a) Determine the oxidation state of oxygen in all the substances. [3]

....

....

....

....

b) Explain how this reaction can be classified as a redox reaction. [3]

....

....

....

....

c) Deduce the oxidising agent. [1]

....

(4) Chlorine reacts with hot sodium hydroxide in the following reaction.

$$6NaOH + 3Cl_2 \longrightarrow 5NaCl + NaClO_3 + 3H_2O$$

a) Give the oxidation state of chlorine in $NaClO_3$ and in $NaCl$. [2]

....

....

b) State, in terms of redox, what happens to chlorine in this reaction. [4]

....

....

....

....

....

Born–Haber cycles

Key words

Lattice formation

Enthalpy of lattice dissociation, $\Delta_{Latt}H$: The energy change when 1 mole of an ionic crystal lattice is broken into its constituent ions in gaseous form. It is always endothermic.

Enthalpy of lattice formation, $\Delta_{Latt}H$: The energy change when 1 mole of an ionic crystal lattice is formed from its constituent ions in gaseous form. It is always exothermic.

The strength of an $\Delta_{Latt}H$ depends on the size and charge of the ions.

Born–Haber cycles

Lattice enthalpies cannot be measured directly and must be calculated using Born–Haber cycles.

Key questions

What are standard conditions?

The theoretical value for $\Delta_{Latt}H$ assumes the perfect ionic model. Therefore, the experimental values derived from Born–Haber will be different because a compound will show covalent character.

Standard conditions are 298 K and 100 kPa.

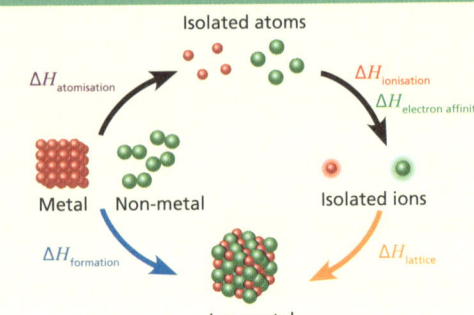

What are the two types of $\Delta_{Latt}H$?

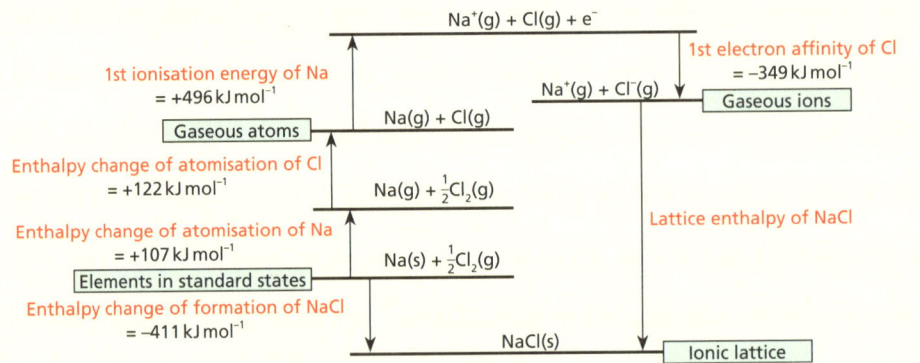

Standard enthalpy of formation, $\Delta_f H^{\ominus}$	The energy transferred when 1 mole of a compound is formed from its elements under standard conditions and all substances are in standard states.
First ionisation energy, $\Delta_{I.E.1}H$	The energy required to remove 1 mole of electrons from 1 mole of gaseous atoms to form 1 mole of gaseous ions with a +1 charge.
Standard enthalpy of atomisation, $\Delta_{at}H$	The energy needed to make 1 mole of gaseous atoms formed from the element in its standard state.
Bond enthalpy, $\Delta_{diss}H$	The energy change when one mole of a covalent bond is broken into two gaseous atoms (free radicals).
What is enthalpy of hydration? / **First electron affinity, $\Delta_{E.A.1}H$**	The energy change when 1 mole of gaseous atoms gains 1 mole of electrons to form 1 mole of gaseous ions with a –1 charge. Subsequent electron affinities will be endothermic owing to the repulsion between the newly added electron and the negative ion.

Enthalpy of hydration

Enthalpy of hydration, $\Delta_{hyd}H$, is the energy change when one mole of gaseous ions is dissolved in water to form one mole of aqueous ions.

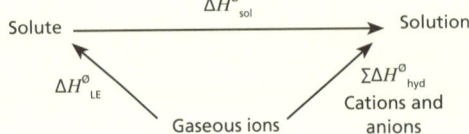

$\Delta_{sol}H = \Delta_{Latt}H + \Sigma\Delta_{hyd}H$
where $\Delta_{Latt}H$ is enthalpy of lattice dissociation

$\Delta_{sol}H = -\Delta_{Latt}H + \Sigma\Delta_{hyd}H$
where $\Delta_{Latt}H$ is enthalpy of lattice formation

Summary

Born–Haber cycles

1 Which equation represents the process that occurs when the standard enthalpy of atomisation of iodine is measured? Tick **one** box. [1]

$Br_2(l) \longrightarrow 2Br(g)$ ☐ $\frac{1}{2}Br_2(l) \longrightarrow Br(g)$ ☐

$\frac{1}{2}Br_2(g) \longrightarrow Br(g)$ ☐ $2Br_2(g) \longrightarrow Br(g)$ ☐

2 Which compound shows the greatest percentage difference between the experimental and theoretical value of the lattice enthalpy? Tick **one** box. [1]

NaCl ☐   CsF ☐   LiI ☐   KF ☐

3 Calcium chloride is an ionic compound that forms when a calcium atom transfers two electrons to two chlorine atoms.

   a) What is meant by 'enthalpy of lattice formation'? [2]

   ....

   ....

   b) Use the data in the table to calculate the enthalpy of lattice dissociation of calcium chloride. [3]

Standard enthalpy change	Value (kJ mol^{-1})
Solution	−82.9
Hydration of calcium ions	−1650
Hydration of chloride ions	−364

   c) Explain why the enthalpy of lattice dissociation for beryllium chloride is different to the enthalpy of lattice dissociation for calcium chloride. [2]

   ....

   ....

4 This question is about the compound silver chloride.

   a) Write an equation for the process of first ionisation energy for silver. [1]

   ....

   b) Name the process illustrated by the equation $Cl(g) + e^- \longrightarrow Cl^-(g)$ [1]

   ....

   c) Complete the Born–Haber cycle for silver chloride by writing the missing species on the dotted lines. [3]

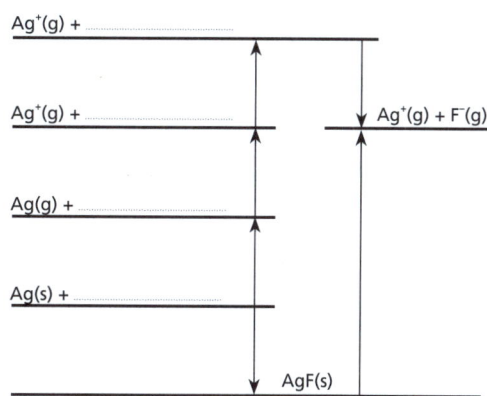

Born–Haber cycles: calculations

Key words

Calculating lattice enthalpy

When drawing a Born–Haber cycle, you must:
- include state symbols (as changing state involves energy)
- ensure the arrows are pointing in the correct direction for the enthalpy change.

Draw a Born–Haber cycle for magnesium chloride and use it to calculate lattice enthalpy.

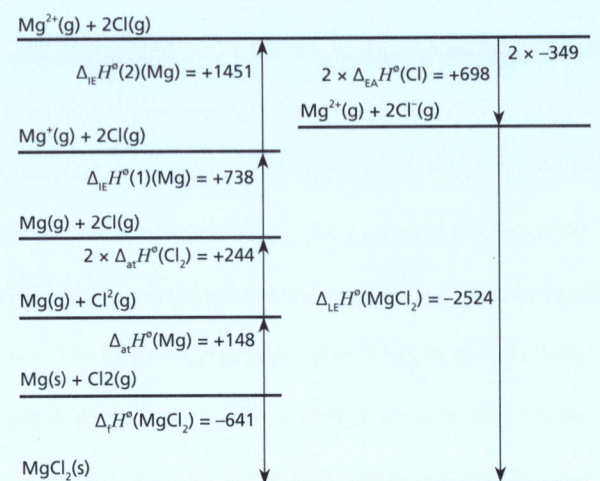

$Mg^{2+}(g) + 2Cl(g)$

$\Delta_{IE}H^{\ominus}(2)(Mg) = +1451$

$2 \times \Delta_{EA}H^{\ominus}(Cl) = +698$

2×-349

$Mg^{2+}(g) + 2Cl^-(g)$

$Mg^+(g) + 2Cl(g)$

$\Delta_{IE}H^{\ominus}(1)(Mg) = +738$

$Mg(g) + 2Cl(g)$

$2 \times \Delta_{at}H^{\ominus}(Cl_2) = +244$

$Mg(g) + Cl^2(g)$

$\Delta_{LE}H^{\ominus}(MgCl_2) = -2524$

$\Delta_{at}H^{\ominus}(Mg) = +148$

$Mg(s) + Cl2(g)$

$\Delta_{f}H^{\ominus}(MgCl_2) = -641$

$MgCl_2(s)$

The lattice enthalpy = −2(first electron affinity of chlorine) − (2nd atomisation enthalpy of Mg) − (1st ionisation enthalpy of Mg) − 2(atomisation enthalpy of Cl) − (atomisation enthalpy of Mg) + (enthalpy of formation of MgCl₂)
$\Delta_{Latt}H = -2(-349) - (+1451) - (+738) - 2(+122) - (+148) + (-641) = -2524 \, kJ \, mol^{-1}$

Key questions

Why should you include state symbols in a Born–Haber cycle?

Why does the direction of the arrow matter in a Born–Haber cycle?

Thermodynamic stability

For substances that have more than one stable ion, the ionic compound with the most exothermic lattice formation enthalpy will be the one that is most thermodynamically stable.

Copper forms stable copper(I), Cu^+ ions, and copper(II), Cu^{2+} ions. Therefore, copper chloride can be copper(I) chloride, CuCl, with a $\Delta_{Latt}H = -996 \, kJ \, mol^{-1}$ or copper(II) chloride, $CuCl_2$, with $\Delta_{Latt}H = -2824 \, kJ \, mol^{-1}$. Of the two, copper(II) chloride is more thermodynamically stable, with stronger ionic bonds, because the copper ion is smaller and more highly charged.

Using Born-Haber cycles

Any single missing value can be calculated from a Born–Haber cycle.

What is 'thermodynamic stability'?

What is the atomisation enthalpy of chlorine?

Atomisation enthalpy = forming one mole of gaseous chlorine atoms from chlorine in its standard state: $0.5Cl_2(g) \longrightarrow Cl(g)$

Find an alternative route between reactants and products using the Born–Haber cycle.

2 × atomisation enthalpy of chlorine = −(atomisation enthalpy of Mg) + (enthalpy of formation of MgCl₂) − (lattice enthalpy) − 2(first electron affinity of chlorine) − (2nd atomisation enthalpy of Mg) − (1st ionisation enthalpy of Mg)

2 × atomisation enthalpy of chlorine = −(+148) + (−641) − (−2524) − 2(−349) − (+1451) − (+738) = 244

$\Delta_{atm}H^{\ominus}(Cl_2) = +122 \, kJ \, mol^{-1}$

Summary

Born–Haber cycles: calculations

1. Which of the following statements is not correct? Tick **one** box. [1]

   Second electron affinity always has a positive enthalpy change ☐

   Ionisation energy always has a positive enthalpy change ☐

   Lattice dissociation enthalpy is always a negative enthalpy change ☐

   Enthalpy of atomisation has the reactants in their standard states ☐

2. $\Delta_{Latt}H^{\ominus}$ for magnesium chloride is $-254\,kJ\,mol^{-1}$

   Which of the following statements is correct? Tick **one** box. [1]

   This is the value of enthalpy of lattice formation ☐

   This is the value of enthalpy of lattice dissociation ☐

   This is the same numerical value as the enthalpy of formation ☐

   The elements magnesium and chlorine are more thermodynamically stable than the compound that they make ☐

3. The Born–Haber cycle for CsCl is shown below.

   $Cs^{+}(g) + Cl(g) + e^{-}$

   $\Delta H_4 = +121\,kJ\,mol^{-1}$ $\Delta H_5 = -364\,kJ\,mol^{-1}$

   $Cs^{+}(g) + \frac{1}{2}Cl_2(g) + e^{-}$ $Cs^{+}(g) + Cl^{-}(g)$

   $\Delta H_3 = +376\,kJ\,mol^{-1}$

   $Cs(g) + \frac{1}{2}Cl_2(g)$

   $\Delta H_2 = +79\,kJ\,mol^{-1}$ ΔH_6

   $Cs^{+}(s) + \frac{1}{2}Cl_2(g)$

   $\Delta H_1 = -433\,kJ\,mol^{-1}$

   $CsCl(s)$

   a) Name the enthalpy changes represented by ΔH_1, ΔH_2, ΔH_3 and ΔH_4. [4]

   ....

   ....

   ....

   b) Why does ΔH_6 have a negative value? [1]

   ....

   c) Calculate a value for the lattice enthalpy. [2]

   ....

4. Draw a Born–Haber cycle for the formation of solid barium chloride, $BaCl_2$. Use your Born–Haber cycle and the standard enthalpy data given below to calculate a value for the electron affinity of chlorine. [8]

Enthalpy of atomisation of barium	$+180\,kJ\,mol^{-1}$
Enthalpy of atomisation of chlorine	$+122\,kJ\,mol^{-1}$
Enthalpy of formation of barium chloride	$-859\,kJ\,mol^{-1}$
First ionisation enthalpy of barium	$+503\,kJ\,mol^{-1}$
Second ionisation enthalpy of barium	$+965\,kJ\,mol^{-1}$
Lattice formation enthalpy of barium chloride	$-2056\,kJ\,mol^{-1}$

Gibbs free-energy change

Entropy

Entropy, S, can be measured directly and has a value in $J\,K^{-1}\,mol^{-1}$.

Entropy is greater when:
- the temperature is higher
- there are a larger number of particles
- there is greater freedom of movement.

Entropy

Entropy is a measure of the degree of disorder of the particles in a thermodynamic system

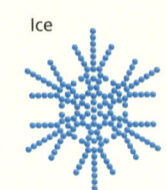

Ice

Ordered
Low entropy

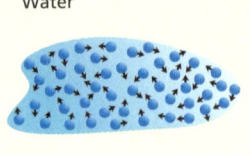

Water

Disordered
High entropy

Key questions

What is entropy?

Feasibility

Reactions that can take place are called feasible or spontaneous. Most chemical reactions are enthalpy driven, have a negative ΔH and are exothermic.

Endothermic reactions are not enthalpy driven as they have a positive ΔH. These reactions are entropy driven and ΔS accounts for this.

What is Gibbs free energy?

Gibbs free energy

For a reaction to be feasible, Gibbs free-energy change, $\Delta G \leqslant 0$.

Convert from °C to K (+273)

Unit of S: $J\,K^{-1}\,mol^{-1}$

$$\Delta G = \Delta H - T\Delta S$$

Units: $kJ\,mol^{-1}$ Units: $kJ\,mol^{-1}$

Need to convert to $kJ\,K^{-1}\,mol^{-1}$ (÷ 1000)

How do you calculate Gibbs free energy?

This expression can be used to determine at what temperature a reaction becomes feasible. It doesn't give any information about the rate of reaction.

Calculate the minimum feasible temperature that the following reaction can occur at.

$$2Al_2O_3(s) + 3C(s) \longrightarrow 4Al(s) + 3CO_2(g)$$

Substance	Al_2O_3(s)	Al(s)	C(s)	CO_2(g)
$\Delta_f H^{\varnothing}$ ($kJ\,mol^{-1}$)	−1669	0	0	−394
$S^{\varnothing}$ ($J\,K^{-1}\,mol^{-1}$)	51	28	6	214

$\Delta H = (3 \times -394) - (-1669 \times 2)$
   $= 2156\,kJ\,mol^{-1}$

$\Delta S = (28 \times 4 + 214 \times 3) - (51 \times 2 + 6 \times 3)$
   $= 634\,J\,K^{-1}\,mol^{-1}$

$\Delta S = 634\,J\,K^{-1}\,mol^{-1} \div 1000$
   $= 0.634\,kJ\,K^{-1}\,mol^{-1}$

$\Delta G = \Delta H - T\Delta S = 0$

$T = \frac{2156}{0.634} = 3401\,K$

 Summary

Gibbs free-energy change

(1) Which one of these has the highest entropy? Tick **one** box. [1]

100 cm³ of ice ☐   100 ml of water at 298 K ☐

100 g of boiling water ☐   100 cm³ of boiling saturated sodium chloride solution ☐

(2) A reaction is exothermic and has a negative entropy change.

Which of the following statements is correct? Tick **one** box. [1]

The reaction is feasible below a certain temperature ☐

The reaction is feasible above a certain temperature ☐

The reaction is always spontaneous ☐

The reaction is never spontaneous ☐

(3) The sketch graph shows how entropy changes with temperature for a pure substance.

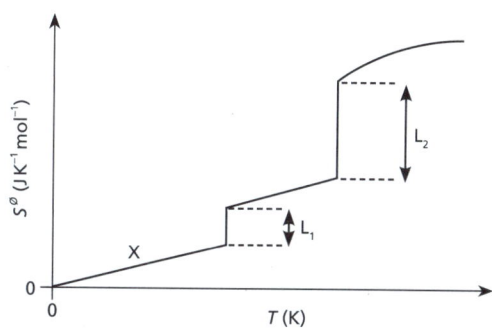

a) Explain why entropy = $0 \, J \, K^{-1} \, mol^{-1}$ at 0 K. [2]

....

....

b) Explain the changes in entropy happening in the first part of the graph from origin to L_1, marked as X. [2]

....

....

c) Compare the changes that occur at L_1 and L_2. [3]

....

....

....

(4) Industrial methanol can be made using the reaction $CO_2(g) + 3H_2(g) \longrightarrow CH_3OH(g) + H_2O(g)$

Draw a Born–Haber cycle for the formation of solid barium chloride, $BaCl_2$. Use the standard enthalpy data and entropy data given below to calculate Gibbs free energy for this reaction at 890 K.

Enthalpy of formation $CO_2(g)$	$-394 \, kJ \, mol^{-1}$
Enthalpy of formation $CH_3OH(g)$	$-201 \, kJ \, mol^{-1}$
Enthalpy of formation of $H_2O(g)$	$-242 \, kJ \, mol^{-1}$
Entropy $CO_2(g)$	$+241 \, J \, K^{-1} \, mol^{-1}$
Entropy $H_2(g)$	$+131 \, J \, K^{-1} \, mol^{-1}$
Entropy $CH_3OH(g)$	$+238 \, J \, K^{-1} \, mol^{-1}$
Entropy $H_2O(g)$	$+189 \, J \, K^{-1} \, mol^{-1}$

[6]

....

Rate equations

Key words

Key questions

What is a rate equation?

What does the rate constant tell you?

What is the Arrhenius equation?

What is the rate determining step?

Rate equation

$$Rate = k[A]^m[B]^n$$

where m and n are the orders of reaction and can be 0, 1 or 2

Overall order of reaction = $m + n$

Zero order	First order	Second order

Concentration vs Time graphs; Rate vs Concentration of A graph labelled 2, 1, 0

The rate constant (k) indicates how fast or slow a reaction occurs, under a given set of conditions. The larger the value of k, the faster the rate of reaction. A mechanism is a series of steps that models a chemical reaction. The species in a rate equation can only be included if one of the following is true:

- The species occur as reactants in the rate determining step.
- The species occur in steps before the rate determining step.

Use the data from the table to determine the rate equation for a reaction   A + B + C ⟶ D

	[A] (mol dm^{-3})	[B] (mol dm^{-3})	[C] (mol dm^{-3})	Rate (dm^{-3} s^{-1})
1	0.40	0.20	0.80	1.60
2	0.40	0.80	0.80	6.40
3	0.80	1.60	0.80	51.2
4	0.80	0.80	0.40	25.6

Compare experiments 1 and 2: first order with respect to B.

Compare experiments 2 and 3: second order with respect to A.

Compare experiments 2 and 4: zero order with respect to C.

Rate = $k[A]^2[B]^1$

Units of rate constant = $mol^{-2} dm^6 s^{-1}$

Arrhenius equation

Rate constant, k, increases with increasing the temperature.

$$k = Ae^{\frac{-E_a}{RT}}$$

k = rate constant
A = pre-exponential factor
E_a = activation energy
R = universal gas constant
T = absolute temperature

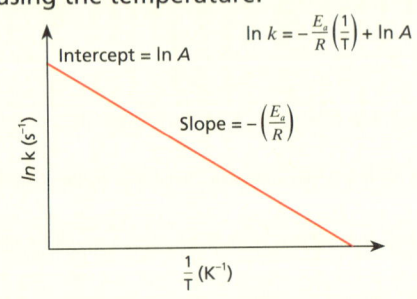

$\ln k = -\dfrac{E_a}{R}\left(\dfrac{1}{T}\right) + \ln A$

Intercept = ln A

Slope = $-\left(\dfrac{E_a}{R}\right)$

$\ln k$ (s^{-1})

$\dfrac{1}{T}$ (K^{-1})

Summary

Rate equations

1 The rate equation for the reaction between substance X and substance Y is   rate = k[X]^2[Y]

Which statement is correct? Tick **one** box.   [1]

When the concentration of X doubles the rate also doubles ⬚

The overall order of reaction is 3 ⬚

Changing the concentration of Y has no effect on the rate ⬚

Rate constant is constant when temperature changes ⬚

2 What are the units of the rate constant in a second order reaction? Tick **one** box.   [1]

No units ⬚

s^{-1} ⬚

$mol^{-1} dm^3 s^{-1}$ ⬚

$mol^{-1} dm^3$ ⬚

3 Methanol can be made from a reaction of a haloalkane with a hydroxide ion.

$CH_3I + OH^- \longrightarrow CH_3OH + I^-$

The initial rate of reaction was studied in three experiments and the results are shown in the table.

	[CH$_3$I] (mol dm^{-3})	[OH]$^-$ (mol dm^{-3})	Rate (mol dm^{-3} s^{-1})
1	0.1	0.1	1 × 10^{-5}
2	0.2	0.1	2 × 10^{-5}
3	0.2	0.4	2 × 10^{-5}

a) Determine the order of reaction with respect to [CH$_3$I]   [1]

b) Determine the order of reaction with respect to [OH$^-$].   [1]

c) Write an expression for the rate equation.   [1]

d) Determine the units of the rate constant.   [1]

4 The Arrhenius equation shows how the rate constant, k, varies with temperature.

Calculate a value for the Arrhenius constant, A, where $k = 3.46 \times 10^{-8} s^{-1}$ at 25°C.

Given that:

$E_a = 96.2 \, kJ \, mol^{-1}$

$R = 8.31 \, J \, K^{-1} \, mol^{-1}$

In your answer, give the units for A.   [4]

Equilibrium constant, K_p, for homogeneous systems

Key words

Key questions

What is a partial pressure?

What is K_p?

When does K_p change?

Partial pressure

Partial pressure of a gas in a mixture is the pressure that the gas would have if it alone occupied the volume occupied by the whole mixture.

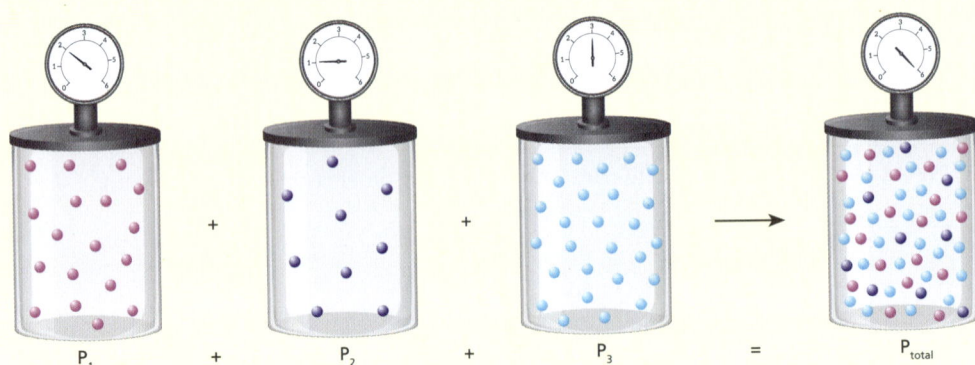

$$P_1 + P_2 + P_3 = P_{total}$$

Where partial pressure of $P_1 = p(P_1)$ (Pa) = mole fraction (mol) × total pressure (Pa)

Equilibrium constant, K_p

The equilibrium constant, K_p, is calculated by finding the product of the partial pressures of the products, and dividing by the product of the partial pressures of the reactants at a constant temperature. This is constant at constant temperature, changing pressure or on addition of a catalyst.

For an equilibrium system: $aA(g) + bB(g) \rightleftharpoons cC(g) + dD(g)$

$$K_p = \frac{p(D)^d \, p(C)^c}{p(A)^a \, p(B)^b}$$

The equilibrium constant, K_p, is used:

- for a homogenous gaseous equilibrium system, e.g. a large K_p implies a greater amount of products; a small K_p suggests equilibrium favours the reactants
- as a measure of how far a chemical reaction reaches at equilibrium.

Methanol can be made from a mixture of hydrogen and carbon monoxide.

$$2H_2(g) + CO(g) \rightleftharpoons CH_3OH(g)$$

The mixture was allowed to reach equilibrium at a fixed temperature. The partial pressure of carbon monoxide was 7550 kPa, the partial pressure of hydrogen was 12 300 kPa and the partial pressure of methanol was 2710 kPa. Calculate the value of K_p.

$$K_p = \frac{pCH_3OH}{pH_2^2 \times pco} = \frac{2710}{(12\,300)^2 \times (7550)} = 2.37 \times 10^{-9}\,kPa^{-2}$$

Summary

Equilibrium constant, K_p, for homogeneous systems

(1) This question relates to the equilibrium gas-phase synthesis of sulfur trioxide:

$2SO_2(g) + O_2(g) \rightleftharpoons 2SO_3(g)$

What are the units for the equilibrium constant, K_p? Tick **one** box. [1]

No units ☐   KPa ☐   Pa ☐ kPa^{-1} ☐

(2) Which of the following statements is correct for the equilibrium constant, K_p? Tick **one** box. [1]

K_p is calculated by mole fraction values for each gas in the equilibrium mixture ☐

Changing the temperature changes the value of K_p ☐

The value of K_p increases when a catalyst is added ☐

The value of K_p is always constant even if conditions change ☐

(3) A mixture of PCl_3 and Cl_2 was heated in a vessel of fixed volume to a constant temperature and the following reaction reached equilibrium.

$PCl_3(g) + Cl_2(g) \rightleftharpoons PCl_5(g)$ $\Delta H = -93\,kJ\,mol^{-1}$

a) Write an expression for the equilibrium constant, K_p. [1]

b) Deduce the units of the equilibrium constant, K_p. [1]

c) Explain the effect of increasing temperature on the value of the equilibrium constant, K_p. [3]

(4) Sulfuryl dichloride, SO_2Cl_2 can undergo thermal decomposition.

$SO_2Cl_2(g) \rightleftharpoons SO_2(g) + Cl_2(g)$ $\Delta H = +93\,kJ\,mol^{-1}$

In one experiment, 1.00 mol of SO_2Cl_2 underwent the reaction. The equilibrium mixture was found to contain 0.75 mol of Cl_2 at 400 K and a total pressure of 125 kPa.

Calculate a value for K_p. [7]

Electrode potentials

Key words

Electrochemical cells generate electrical energy from chemical reactions

In an electrochemical cell:

- redox reactions occur
- electrons are transferred from the reducing agent to the oxidising agent indirectly via an external circuit
- potential difference is produced, which can drive an electric current to do work.

Each electrochemical cell is made from a half-cell, which is usually a metal in contact with a solution of its ions. The half-cells are connected with a salt bridge, which completes the circuit and carries the charge with free moving ions.

Key questions

What is an electrochemical cell?

A conventional diagram can be used to represent this cell:

Where you read from left to right to generate the half-equations.

Reaction at the anode: $Zn \longrightarrow Zn^{2+} + 2e^-$

Reaction at the cathode: $Cu^{2+} + 2e^- \longrightarrow Cu$

Then the ionic equation would be:

$Zn + Cu^{2+} \longrightarrow Zn^{2+} + Cu$

The electrons are donated from the zinc atom to the copper ion.

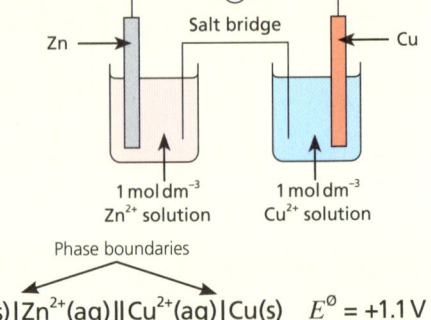

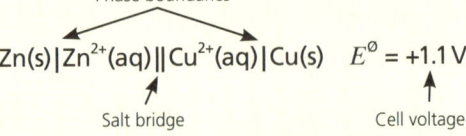

$$Zn(s)|Zn^{2+}(aq)\|Cu^{2+}(aq)|Cu(s) \quad E^{\varnothing} = +1.1\,V$$

Salt bridge — Cell voltage

Standard electrode potentials

What is SHE?

The electrode potential of a half-cell is measured in reference to a standard hydrogen electrode (SHE) under standard conditions of 298 K, 100 kPa and all solutions at a concentration of $1\,mol\,dm^{-3}$. Standard electrode potentials can be listed as an electrochemical series.

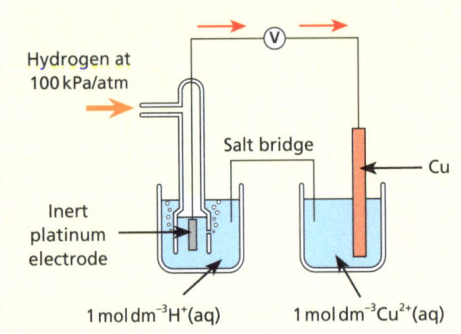

What are the standard conditions?

Oxidising agent		Reducing agent	Standard potential $E^{\varnothing}$ (V)
$K^+(aq)$	$+ e^- \rightleftharpoons$	$K(s)$	−2.92
$Mg^{2+}(aq)$	$+ 2e^- \rightleftharpoons$	$Mg(s)$	−2.38
$Zn^{2+}(aq)$	$+ 2e^- \rightleftharpoons$	$Zn(s)$	−0.76
$Fe^{3+}(aq)$	$+ 3e^- \rightleftharpoons$	$Fe(s)$	−0.04
$2H^+(aq)$	$+ 2e^- \rightleftharpoons$	$H_2(g)$	0.00
$Cu^{2+}(aq)$	$+ 2e^- \rightleftharpoons$	$Cu(s)$	+0.34
$Fe^{3+}(aq)$	$+ e^- \rightleftharpoons$	$Fe^{2+}(aq)$	+0.77
$Ag^+(aq)$	$+ e^- \rightleftharpoons$	$Ag(s)$	+0.80
$O_2(g) + 4H^+(aq)$	$+ 2e^- \rightleftharpoons$	$2H_2O(l)$	+1.23
$MnO_4^-(aq) + 8H^+(aq)$	$+5e^- \rightleftharpoons$	$Mn^{2+}(aq) + 4H_2O$	+1.49

Calculate the cell potential, $E^{\varnothing}$ cell (electromotive force or EMF), of the electrochemical cell $Zn|Zn^{2+}\|Cu^{2+}|Cu$.

$Cu^{2+}(aq) + 2e^- \rightleftharpoons Cu(s) \qquad E^{\varnothing} = +0.34\,V$

$Zn^{2+}(aq) + 2e^- \rightleftharpoons Zn(s) \qquad E^{\varnothing} = -0.76\,V$

$E^{\varnothing}_{cell} = E^{\varnothing}_{Right} - E^{\varnothing}_{Right}$

$E^{\varnothing}_{cell} = +0.34 - (-0.76) = +1.10\,V$

The half-cell with the most negative $E^{\varnothing}$ will donate electrons to the half-cell with the most positive $E^{\varnothing}$.

✔ Summary

Electrode potentials

(1) What is the $E^{\varnothing}$ cell for the reaction between zinc and magnesium half-cells? Tick **one** box. [1]

$Mg^{2+}(aq) + 2e^- \rightleftharpoons Mg(s) \quad E^{\varnothing} = -2.38$ V

$Zn^{2+}(aq) + 2e^- \rightleftharpoons Zn(s) \quad E^{\varnothing} = -0.76$ V

+1.62 V ☐   −1.62 V ☐   3.14 V ☐   −3.14V ☐

(2) Which of the following statements is correct about a salt bridge in an electrochemical cell? Tick **one** box. [1]

The salt bridge allows electrons to flow ☐

A salt bridge completes the circuit ☐

The salts used in the salt bridge must be in crystalline form ☐

Salt bridges act as a catalyst by lowering the activation energy of the redox reaction ☐

(3) An electrochemical cell was set up under standard conditions. The conventional cell diagram for this cell is:

$Pt(s) \mid H_2(g) \mid 2H^+(aq) \parallel Fe^{3+}(aq), Fe^{2+}(aq) \mid Pt(s)$

a) List the standard conditions. [3]

_____

_____

b) Name the half-cell which forms the anode. [1]

_____

c) Write a half-equation for the reaction at the cathode. [1]

_____

d) Describe the movement of electrons in this cell. [3]

_____

_____

_____

(4) An electrochemical cell can be made from a copper half-cell attached to a silver half-cell by a salt bridge.

The standard electrode potentials are:

$Cu^{2+}(aq) + 2e^- \rightleftharpoons Cu(s) \quad 0.34$ V

$Ag^+(aq) + e^- \rightleftharpoons Ag(s) \quad 0.80$ V

a) Draw a conventional cell diagram for this cell. [3]

b) Write an ionic equation to represent the reaction occurring in this electrochemical cell. [2]

_____

c) Calculate the electromotive force (EMF) for this electrochemical cell. [2]

Commercial cells

Key words

Key questions

What happens in a non-rechargeable cell?

How do you recharge a rechargeable cell?

What is a fuel cell?

What are the advantages of fuel cells over electrochemical cells?

Use of commercial cells

Electrochemical cells can be used as a commercial source of electrical energy. They can be a portable supply of electricity to power electronic devices (e.g. mobile phones) and they can provide energy to power a vehicle.

Non-rechargeable (primary) cells

In a non-rechargeable cell, the redox reaction is non-reversible. Non-rechargeable cells are cheap to manufacture and expensive to recycle.

A Leclanché cell, or dry cell, is an example of a non-rechargeable cell.

Anode: $Zn \longrightarrow Zn^{2+} + 2e^-$

$E^{\ominus} = -0.80$ V

Cathode: $2NH_4^+(aq) + 2e^- \longrightarrow 2NH_3(g) + H_2(g)$

$E^{\ominus} = +0.40$ V

$E^{\ominus}_{cell} = E^{\ominus}_{Right} - E^{\ominus}_{Right}$

$E^{\ominus}_{cell} = +0.40 - (-0.80) = +1.50$ V

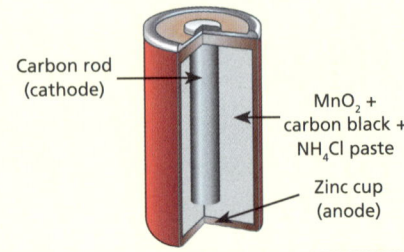

Carbon rod (cathode)

MnO_2 + carbon black + NH_4Cl paste

Zinc cup (anode)

Rechargeable (secondary) cells

In a rechargeable cell, the redox reaction is reversible when an external voltage is applied greater than the cell voltage. Rechargeable cells are more expensive to manufacture and recycle than non-rechargeable cells, but can be re-used many times.

A lithium cell is an example of a rechargeable cell.

$Li \parallel Li^+ \mid Li, CoO_2 \mid LiCoO_2 \mid Pt$

Anode: $Li^+ + CoO_2 + e^- \longrightarrow Li^+ [CoO_2]^-$

$E^{\ominus} = +0.90$ V

Cathode: $Li \longrightarrow Li^+ + e^-$

$E^{\ominus} = -3.04$ V

$E^{\ominus}_{cell} = E^{\ominus}_{Right} - E^{\ominus}_{Right}$

$E^{\ominus}_{cell} = +0.90 - (-3.04) = +3.94$ V

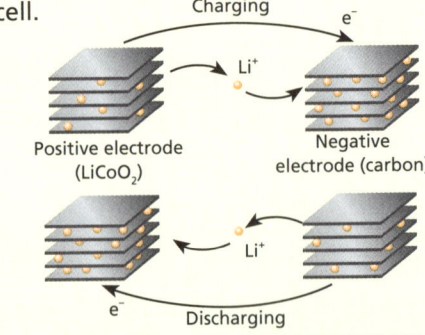

Charging

e^-

Li^+

Positive electrode ($LiCoO_2$)

Negative electrode (carbon)

Li^+

e^-

Discharging

Fuel cells

Fuel cells are used to generate an electric current and do not need to be recharged.

An alkaline hydrogen–oxygen fuel cell is an example of a fuel cell.

This fuel cell uses hydrogen and oxygen to produce electricity, heat and potable water.

Anode: $O_2 + 2H_2O + 4e^- \longrightarrow 4OH^-$ (reduction)

Cathode: $2H_2 + 4OH^- \longrightarrow 4H_2O + 4e^-$ (oxidation)

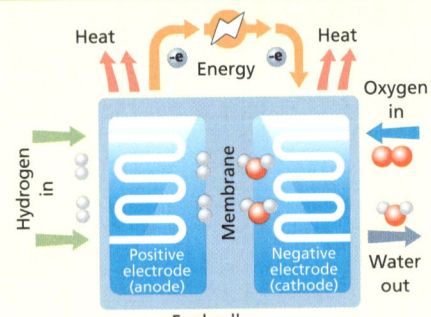

Heat

Energy

Heat

Oxygen in

Hydrogen in

Membrane

Positive electrode (anode)

Negative electrode (cathode)

Water out

Fuel cell

Advantages of hydrogen fuel cells	Disadvantages of hydrogen fuel cells
• Quiet, efficient and small in size • Easy to maintain as they have no moving parts • Make no pollution from the products • Sustainable, as hydrogen can be made from the electrolysis of water using solar power	• Expensive to manufacture • Hydrogen fuel is a flammable gas, which is difficult to store and can be dangerous to store and use

Summary

Spec. ref. 3.1.11.2

Commercial cells

1. Lithium-ion cells are used to power mobile phones. The conventional representation of this cell is shown below.

   $Li \mid Li^+ \parallel Li^+, CoO_2 \mid LiCoO_2 \mid Pt$

   Which element undergoes a change in oxidation state at the anode? Tick **one** box. [1]

   Li ☐   Co ☐   O ☐   Pt ☐

2. Nickel–cadmium cells are used to power electric shavers. The electrode reactions are shown below.

   $NiO(OH) + H_2O + e^- \longrightarrow Ni(OH)_2 + OH^- \quad E^{\varnothing} = +0.52$ V

   $Cd(OH)_2 + 2e^- \longrightarrow Cd + 2OH^- \quad E^{\varnothing} = -0.88$ V

   What is the EMF of a nickel–cadmium cell? Tick **one** box. [1]

   +1.40 V ☐

   −1.40 V ☐

   −0.36 V ☐

   +0.36 V ☐

3. Lithium cells are used to power tablet computers. The conventional representation of a lithium cell is given below.

   $Li(s) \mid Li^+(aq) \parallel Li^+(aq) \mid MnO_2(s), LiMnO_2(s) \mid Pt\ (s)$

   The cell has an EMF of +2.91 V.

   a) Write a half-equation for the reaction that occurs at the positive electrode of this cell. [2]

   b) Write a half-equation for the reaction that occurs at the negative electrode of this cell. [2]

   c) The standard electrode potential, $E^{\varnothing}$, for the negative electrode is −3.04V.

   Calculate the standard electrode potential, $E^{\varnothing}$, for the positive electrode. [1]

4. A representation of an alkaline hydrogen–oxygen fuel cell is shown below.

   $Pt \mid H_2 \mid H_2O \parallel O_2 \mid OH^- \mid Pt$

   a) Write a half-equation for the reaction that occurs at each electrode. [2]

   b) Deduce the overall equation for the reaction in the hydrogen fuel cell. [1]

   c) Give the main advantage of using a hydrogen–oxygen fuel cell to power a car. [1]

Acids, pH and K_w

🔑 Key words

❓ Key questions

What is an acid?

What is a base?

What is pH?

What is K_w?

Acid–base definitions

Brønsted–Lowry definitions:

- Acid is a proton donor.
- Base is a proton acceptor.
- Acid–base equilibria involve the transfer of protons.

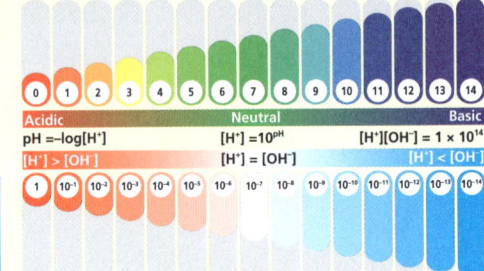

$H_2SO_4(aq)$
Brønsted–Lowry acid

$H_2O(l)$
Brønsted–Lowry base

$HSO_4^-(aq)$
Brønsted–Lowry base

$H_3O^+(aq)$
Brønsted–Lowry acid

Conjugate acid-base pair
H_2SO_4 / HSO_4^-

Conjugate acid-base pair
H_2O / H_3O^+

pH scale

The pH scale is a measure of the concentration of hydrogen ions (protons) in solution. There is a very wide range so a logarithmic scale is used: $pH = -\log_{10}[H^+]$

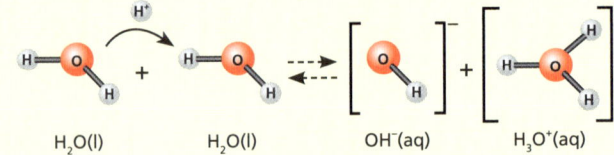

Acidic   Neutral   **Basic**
$pH = -\log[H^+]$ $[H^+] = 10^{pH}$ $[H^+][OH^-] = 1 \times 10^{14}$
$[H^+] > [OH^-]$ $[H^+] = [OH^-]$ $[H^+] < [OH^-]$

Calculate the pH of $0.1\,mol\,dm^{-3}$ nitric acid, HNO_3

$pH = -\log_{10}[H^+]$

$pH = -\log_{10}[0.1] = 1$

> Nitric acid is a strong monoprotic acid and will fully ionise in solution. So, $[HNO_3] = [H^+]$

Ionic product of water, K_w

In any sample of water, some of the water molecules will have dissociated in an endothermic reaction.

$H_2O(l)$ $H_2O(l)$ $OH^-(aq)$ $H_3O^+(aq)$

This is a reversible reaction but because the concentration of water is so large and therefore almost constant, K_c is not meaningful. So, the ionic product of water, K_w is used: $K_w = [H^+][OH^-]$

Calculate the pH of pure water at 298 K where $K_w = 1 \times 10^{-14}\,mol^2\,dm^{-6}$

In pure water, $[H^+] = [OH^-]$

$[H^+] = \sqrt{1 \times 10^{-14}} = 10^{-7}$

$pH = -\log_{10}[H^+] = pH = -\log_{10}[10^{-7}] = 7$

> Water is always neutral as $[H^+] = [OH^-]$ but because the value of Kw varies with temperature, pH also varies with temperature.

Calculate the pH of $0.1\,mol\,dm^{-3}$ potassium hydroxide, KOH, given that $K_w = 1 \times 10^{-14}\,mol^2\,dm^{-6}$

$1 \times 10{-14} = [H^+][0.1]$

$[H^+] = 1 \times 10{-14} \div 0.1 = 1 \times 10^{-13}$

$pH = -\log_{10}[H^+] = -\log_{10}[1 \times 10^{-13}] = 13$

> Potassium hydroxide is a strong alkali and will fully ionise in solution. So, $[KOH] = [OH^-]$

✔ Summary

Acids, pH and K_w

1 Which species can behave as a Brønsted–Lowry acid in aqueous solution? Tick **one** box. [1]

H_3O^+ ☐ NH_3 ☐ SO_4^{2-} ☐ H_2O ☐

2 A strong diprotic acid has a formula H_2X.

What is the pH of a sample of this acid with a concentration of $0.02\,mol\,dm^{-3}$? Tick **one** box. [1]

1.40 ☐

2.78 ☐

−1.40 ☐

1.0 ☐

3 Pure water dissociates slightly.

$H_2O(l) \rightleftharpoons H^+(aq) + OH^-(aq)$ $\Delta H = +57\ kJ\,mol^{-1}$

a) Write an expression for the equilibrium constant for this reaction. [1]

....

b) Explain why $[H_2O]$ is not used in the expression for the ionic product of water. [1]

....

....

c) Explain how the ionic product of water varies with temperature. [3]

....

....

....

....

d) Explain why the pH of water is always 7. [1]

....

....

4 Nitric acid (HNO_3) is a strong acid. Ethanoic acid (CH_3COOH) is a weak acid.

Ethanoic acid can react as a base with pure nitric acid.

a) State the meaning of the term Brønsted–Lowry base. [1]

....

b) Write an equation for the reaction between pure ethanoic acid and pure nitric acid. [1]

....

c) A solution of nitric acid is made with a pH of 1.35.

Calculate the concentration of the acid. [3]

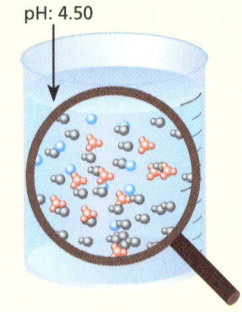

Weak acids

Key words

Key questions

What is a weak acid?

How do you calculate K_a?

What does K_a tell you?

Weak acid

Weak acids:

- partially ionise in aqueous solution
- K_a release $H^+(aq)$ in solution.

For example, for methanoic acid: $HCOOH \rightleftharpoons HCOO^- + H^+$

A general equation for the ionisation of a weak acid is

$$HA \rightleftharpoons A^- + H^+$$

The expression for the acid dissociation constant, K_a for the weak acid HA is:

$$K_a = \frac{[H^+][A^-]}{[HA]}$$

pH: 4.50

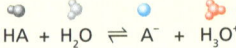

$HA + H_2O \rightleftharpoons A^- + H_3O^+$

> A $0.150\,mol\,dm^{-3}$ sample of a weak acid, HX, was found to have 1.80% dissociation at 25°C. Calculate the value of K_a, giving the units in your answer.
>
> $[H^+] = (1.80 \times 10^{-2}) \times 0.150 = 2.70 \times 10^{-3}$
>
> $K_a = \frac{[H^+][X-]}{[HX]} \approx K_a = \frac{[H^+]^2}{[HX]} = \frac{(2.70 \times 10-3)^2}{0.150} = 4.86 \times 10^{-5}\,mol\,dm^{-3}$
>
> As HA forms H^+ and A^- we can say that $[H^+] = [A^-] = [H^+]^2$

pK_a

K_a indicates the strength of the acid. The higher the number, the stronger the acid. K_a is often small and has a large range, so pK_a is often used as it has a narrower range and is easier to use.

$$pK_a = -\log 10 K_a$$

The lower the pK_a value, the stronger the acid.

> **Propanoic acid is a weak acid with a pK_a value of 4.87. What is the value of K_a for propanoic acid?**
>
> $pK_a = 4.87 = -\log 10 K_a$
>
> $K_a = 10 - pK_a = 1.35 \times 10^{-5}\,mol\,dm^{-3}$

Weak bases

Weak bases (alkalis):

- partially ionise in aqueous solution
- release OH^- (aq) ions in solution.

For example, for ammonium hydroxide:

$NH_4OH \rightleftharpoons NH_3 + OH^-$

pH: 9.50

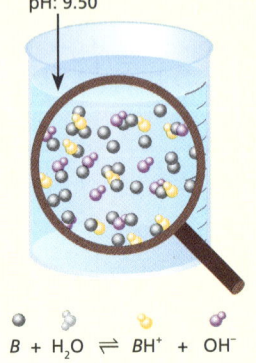

$B + H_2O \rightleftharpoons BH^+ + OH^-$

Summary

Weak acids

1. What is the pH of a $4.25 \times 10^{-3}\,mol\,dm^{-3}$ solution of a weak acid, HA? The acid dissociation constant, K_a, of HA has the value $2.56 \times 10^{-4}\,mol\,dm^{-3}$

   Tick **one** box. [1]

   2.98 ☐

   2.56 ☐

   2.00 ☐

   5.96 ☐

2. The table shows the pK_a values for two acids.

Name of acid	pK_a
Hydrocyanic acid	9.3
Nitrous acid	3.3

   Which of the following statements is correct? Tick **one** box. [1]

   Hydrocyanic acid is a weaker acid than nitrous acid ☐

   Nitrous acid has a greater degree of ionisation than hydrocyanic acid ☐

   Samples of both acids at the same concentration would have the same pH value ☐

   The value of K_a for hydrocyanic acid is greater than that for nitrous acid ☐

3. Ethanoic acid is a weak acid and has K_a value of $1.75 \times 10^{-5}\,mol\,dm^{-3}$ at 25°C.

   $$CH_3COOH\,(aq) \rightleftharpoons H^+\,(aq) + CH_3COO^-\,(aq)$$

   a) Write an expression for the acid dissociation constant for ethanoic acid. [1]

   b) Calculate the concentration of ethanoic acid in a solution of the acid that has a pH of 2.69 at 25°C. [4]

   c) Chloroethanoic acid, $ClCH_2COOH$, is a weak acid with a K_a value of $1.38 \times 10^{-3}\,mol\,dm^{-3}$ at 25°C.

   Compare the strength of ethanoic acid and chloroethanoic acid. [3]

4. A solution of a weak acid has a pH value of 2.34

   A $100\,cm^3$ sample of $0.150\,mol\,dm^{-3}$ solution of this weak acid was studied.

   Calculate the value of pK_a. Give your answer to 2 decimal places. [5]

Titrations

Key words

Key questions

How do you choose a suitable indicator?

What is a titration curve?

What is the relationship between the end-point and the equivalence point?

Titrations

Titrations are an analytical technique where the quantity of an unknown concentration of a substance is reacted fully with a known reagent. The end-point is when the indicator changes colour; this is a good approximation of the equivalence point.

> An aqueous solution of sodium hydroxide is added from a burette to an aqueous solution of $25.00\,cm^3$ of $0.410\,mol\,dm^{-3}$ ethanoic acid. The end-point is reached when $22.60\,cm^3$ of sodium hydroxide solution has been added. Calculate the concentration of sodium hydroxide solution used.
>
> $$CH_3COOH + NaOH \longrightarrow CH_3COONa + H_2O$$
>
> amount of substance (mol) = concentration ($mol\,dm^{-3}$) × volume (dm^3)
>
> amount of acid (mol) = $25.0 \times 10^{-3} \times 0.41 = 1.025 \times 10^{-2}$ = amount of NaOH (mol)
>
> $[NaOH] = \frac{1.025 \times 10^{-2}}{22.6 \times 10^{-3}} = 0.454\,mol\,dm^{-3}$

Indicators

Acid–base indicators are weak acids. The acid and conjugate base are different colours.

Acid–base indicators have an obvious colour change.

It is important to choose an indicator that will clearly change colour in the pH range of the equivalence point. Titrations between a weak acid and a weak base have no suitable indicator as the reaction is slow and there is no sharp change in pH.

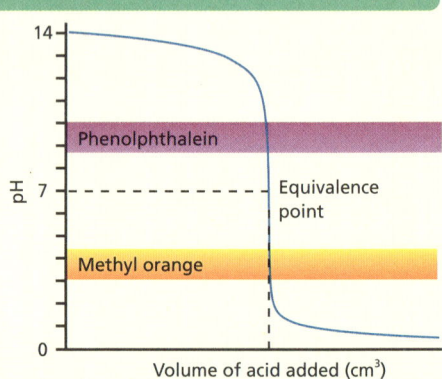

pH curves

The pH can be monitored constantly using a pH probe. When the probe is connected to a storage device and the data is stored, it is called a datalogger. The data can be used to plot pH curves.

The equivalence point is needed for calculations. This is where the acid and the base are equimolar, and it occurs at the midpoint of the steepest part of the curve. For a titration involving a weak acid, half the equivalence point is where the [weak acid] = [conjugative base].

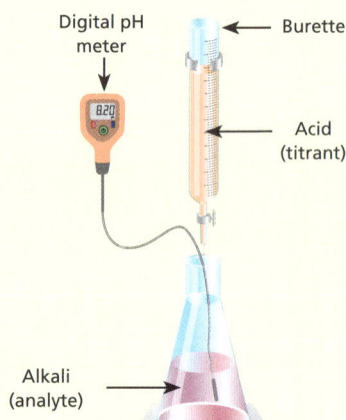

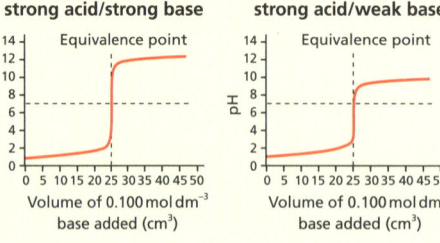

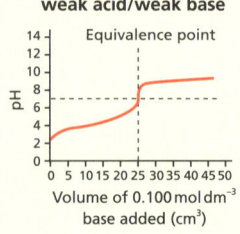

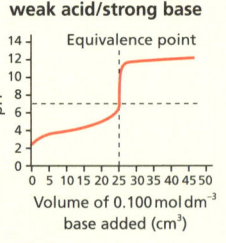

Summary

Titrations

1. Which indicator should be used in a titration to find the concentration of a solution of ammonium hydroxide using $0.010\,mol\,dm^{-3}$ nitric acid? Tick **one** box. [1]

Thymol blue (pH range 1.2–2.8) ☐   Bromophenol blue (pH range 3.0–4.6) ☐

Phenol red (pH range 6.8–8.4) ☐   Phenolphthalein (pH range 8.3–10.0) ☐

2. Using a pipette, exactly $25.00\,cm^3$ of a $0.10\,mol\,dm^{-3}$ aqueous solution of a base is transferred into a conical flask. A $0.10\,mol\,dm^{-3}$ aqueous solution of an acid is slowly added and the pH is continuously monitored.

Which acid–base pair has the highest pH at the equivalence point? Tick **one** box. [1]

$CH_3CH_2COOH + KOH$ ☐ $CH_3CH_2COOH + NH_3$ ☐ $HCl + KOH$ ☐ $HCl + NH_3$ ☐

3. The pH was monitored continuously as a solution of sodium hydroxide solution was added gradually from a burette to $25\,cm^3$ of $0.080\,mol\,dm^{-3}$ propanoic acid under standard conditions. The pH curve shows the results of the investigation.

a) Determine the value of K_a for propanoic acid at 25°C. [4]

b) Methyl orange is an acid–base indicator with a pH range of 3.1–4.4

Evaluate the use of methyl orange in this investigation. [3]

4. A pH curve for an acid–base titration is shown.

a) Give the formula of a suitable acid and the formula of a suitable base that could have produced this pH curve. [2]

b) Suggest how to practically determine which is the best indicator for this reaction. [2]

Buffers

Key words

Key questions

What is a buffer?

How can you make a buffer?

When do you use buffers?

Buffer solutions

A buffer resists changes in pH when:

- a small amount of acid or base is added
- it is diluted with water.

$$CH_3COOH(aq) + H_2O(l) \rightleftharpoons H_3O^+(aq) + CH_3COO^-(aq)$$

H$^+$ Addition   OH$^-$ Addition

Buffer solution after acid addition

Buffer solution

Buffer solution after base addition

Acidic buffers and basic buffers

Acidic buffer solutions contain a weak acid and the salt of that weak acid, where there is a much higher concentration of the salt than the weak acid.

Basic buffer solutions contain a weak base and the salt of that weak base.

> What is the pH of a buffer made from 0.12 mol dm^{-3} ethanoic acid and 0.10 mol dm^{-3} sodium ethanoate?
>
> K_a for ethanoic acid = 1.7 × 10−5 mol dm^{-3}
>
> $[H^+]$ = (1.7 × 10^{-5}) × 0.12 ÷ 0.1 = 2.04 × 10^{-5} mol dm^{-3}
>
> pH = −log 2.04 × 10^{-5} = 4.7

Making and using buffers

An acidic buffer solution can be made by:

- adding the correct mass of salt to a weak acid

  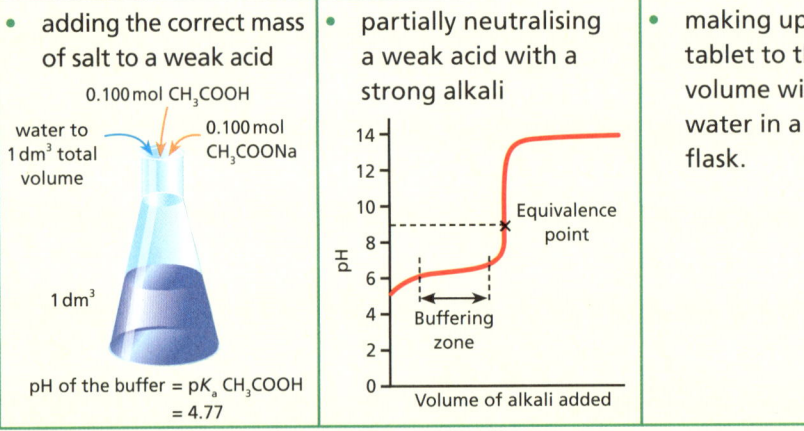

  0.100 mol CH$_3$COOH

  water to 1 dm^3 total volume

  0.100 mol CH$_3$COONa

  1 dm^3

  pH of the buffer = pK_a CH$_3$COOH = 4.77

- partially neutralising a weak acid with a strong alkali

  Equivalence point

  Buffering zone

  Volume of alkali added

- making up a buffer tablet to the correct volume with distilled water in a volumetric flask.

Important applications of buffers:

- Industry, e.g. fabric dying to maintain the pH of dyes so they work effectively
- Biological systems, e.g. buffers naturally occurring in blood to maintain pH ≈ 7.4

> **Mixing together 100 cm^3 of 0.25 mol dm^{-3} ethanoic acid with 100 cm^3 of 0.10 mol dm^{-3} sodium hydroxide:**
>
> moles CH$_3$COOH = 0.25 × 100 × 10^{-3} = 2.5 × 10^{-2} moles $\boxed{\text{moles = concentration × volume}}$
>
> moles NaOH = 0.1 × 100 × 10^{-3} = 1.0 × 10^{-2} moles
>
> The acid is in excess so all of the NaOH will react to form the salt. 1 mole NaOH forms 1 mole salt.
>
> Moles of salt = starting moles of NaOH = 1.0 × 10^{-2} moles
>
> Some acid is used up in making salt. 1 mole acid forms 1 mole salt.
>
> Moles of CH$_3$COONa remaining = 2.5 × 10^{-2} − 1.0 × 10^{-2} = 1.5 × 10^{-2} moles
>
> Total volume of solution = volume of acid + volume of base = 100 + 100 = 200 cm^3
>
> Concentration of acid = moles of acid ÷ total volume in dm^3
>
> Concentration of acid = 1.5 × 10^{-2} ÷ 200 × 10^{-3} = 0.075 mol dm^{-3}
>
> Concentration of salt = 1.0 × 10^{-2} ÷ 200 × 10^{-3} = 0.050 mol dm^{-3}
>
> $[H^+] = K_a \times \frac{\text{Acid}}{\text{Salt}}$ = 1.7 × 10^{-5} $\frac{0.075}{0.050}$ = 2.55 × 10^{-5} mol dm^{-3}
>
> pH = −log[H$^+$] = −log 2.55 × 10^{-5} = 4.6
>
> pH of the buffer formed is pH 4.6.

Summary

Buffers

1. Equal volumes of pairs of solutions are mixed.

   Which pair forms a buffer solution? Tick **one** box. [1]

   Ammonia and ammonium hydroxide ☐   Ethanoic acid and sodium ethanoate ☐

   Ethanoic acid and ammonium hydroxide ☐   Ethanoic acid and propanoic acid ☐

2. A titration curve for a weak acid and strong base is shown.

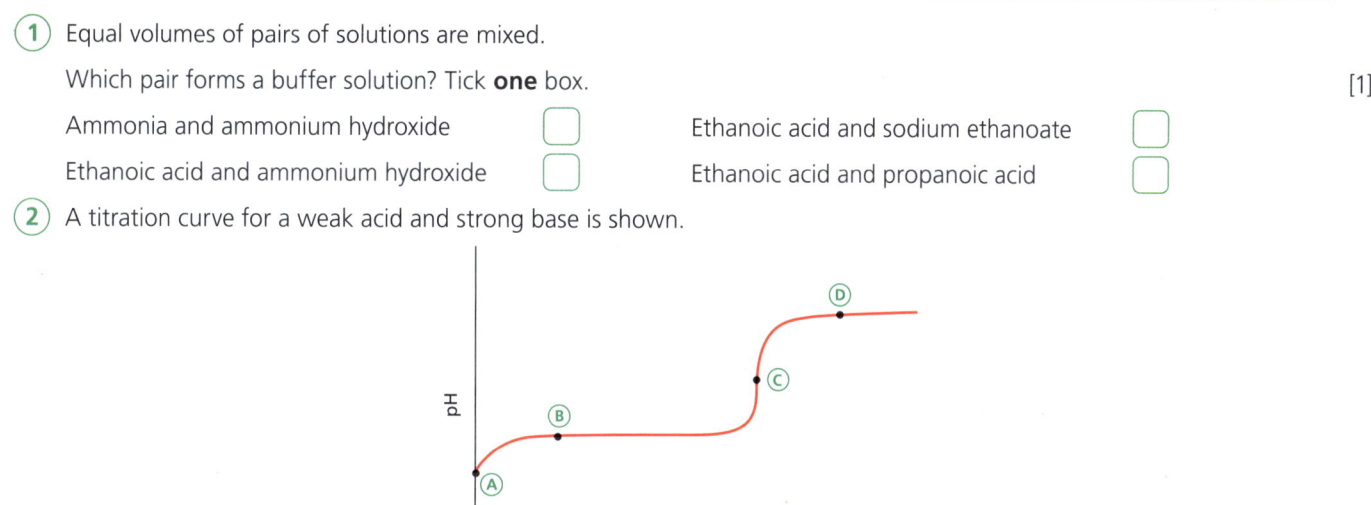

   Where on the titration curve is the solution acting as a buffer solution? Tick **one** box. [1]

   A ☐   B ☐   C ☐   D ☐

3. Water in a swimming pool is kept at a constant pH by using a buffer. This is done by adding sodium hydrogencarbonate and sodium carbonate.

   Hydrogen carbonate ions (HCO_3^-) act as a weak acid in aqueous solution according to the equation $HCO_3^- \rightleftharpoons CO_3^{2-} + H^+$

   a) Describe the conditions in which a buffer resists changes in pH. [3]

   b) Use a mathematical expression to show how a buffer resists changes in pH. [1]

   c) Use the equilibrium equation to explain how a solution containing sodium hydrogencarbonate and sodium carbonate can act as a buffer when small amounts of acid or small amounts of alkali are added. [4]

4. An acidic buffer solution is made by dissolving sodium propanoate (C_2H_5COONa) in a solution of propanoic acid. The mixture is made up to $500\,cm^3$ and the acidic buffer has a pH of 4 with the final concentration of propanoic acid being $0.250\,mol\,dm^{-3}$.

   Calculate the mass of sodium propanoate used. For propanoic acid, $K_a = 1.35 \times 10^{-5}\,mol\,dm^{-3}$. [6]

Key words

Key questions

How are elements grouped into blocks in the Periodic Table?

What is the general trend in atomic radius and ionisation energy as you go across Period 3?

What affects the melting point trends in Period 3?

The Periodic Table and periodicity

The Periodic Table is a list of 118 known elements in ascending atomic (proton) number. Elements can be classified into blocks by their position in the Periodic Table. The blocks are named after the sub-shell that contains the highest energy electron for those elements.

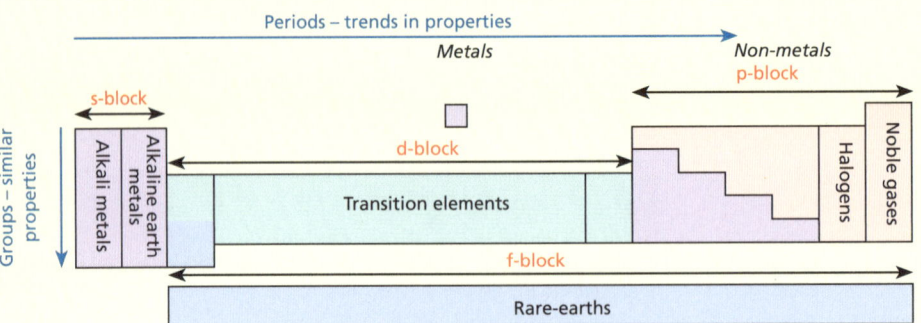

Periodicity is the repeating pattern or trends of properties of the elements in the Periodic Table.

As you go across Period 4, effective nuclear charge increases and shielding remains almost constant.

Therefore, the atomic radius decreases and the first ionisation energy increases because the electrons are more tightly bound to the nucleus.

The first ionisation energy tells you about the electronic sub-shells. The filled and half-filled sub-shells have higher first ionisation because they are energetically more stable. See page 8 for more details.

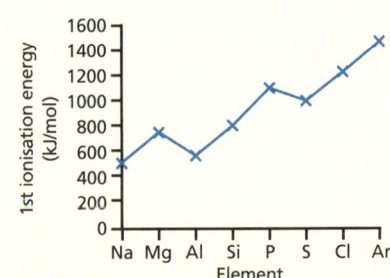

Melting point

The melting point is related to the strength of the forces between the species that need to be overcome to change state.

Period 3	Na	Mg	Al	Si	P	S	Cl	Ar
Structure	giant metallic			giant covalent	P_4	S_8	Cl_2	atomic
					simple covalent			

In giant structures, many strong bonds must be broken to melt the substance and this takes a lot of energy:

- Sodium, magnesium and aluminium have metallic bonds and they increase in strength across the period because the ions get smaller and there are more outer-shell electrons that can be delocalised.
- Silicon is a giant covalent molecule (macromolecule) and many strong covalent bonds (shared pairs of electrons) must be broken.

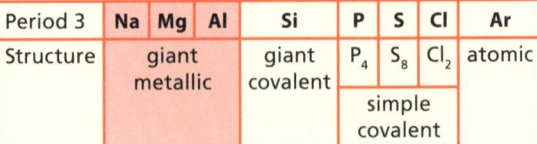

In simple molecules, relatively weak, induced dipole-dipole forces need to be overcome:

- The larger the molecule, the more easily polarisable the electrons will be and the stronger the dipole-dipole forces, and therefore the higher the melting point.

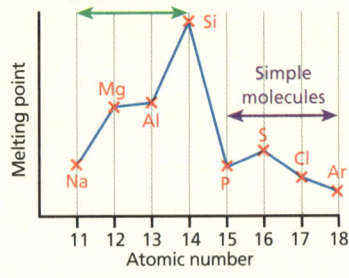

Summary

Periodicity

1. Which statement is not correct about the Period 3 elements? Tick **one** box. [1]

Sodium has the largest atomic radius ☐

Silicon has the highest melting point ☐

All Period 3 elements are in the same block ☐

Phosphorus has a higher first ionisation energy than aluminium ☐

2. Which block in the Periodic Table contains the element praseodymium (Pr)? Tick **one** box. [1]

s-block ☐

p-block ☐

d-block ☐

f-block ☐

3. This question is about Period 3 elements.

   a) Explain the general trend in first ionisation energy as you move across Period 3. [4]

   b) Explain why aluminium deviates from the general trend in first ionisation energies across Period 3. [2]

4. This question is about the melting points of Period 3 elements.

   a) Which element in Period 3 has the highest melting point? [1]

   b) Explain why the melting point of sulfur (S_8) is greater than that of phosphorus (P_4). [3]

Group 2

Key words

Key questions

Where are the alkaline earth metals found in the Periodic Table?

Group 2

Alkaline earth metals are found in the s-block and second column of the Periodic Table. The elements have metallic bonds and form giant lattice structures.

Group 2 element	Reaction with oxygen	Reaction with water
Beryllium	Reluctant to burn; white flame	No reaction
Magnesium	Burns easily; bright white flame	Vigorous with $H_2O(g)$; no reaction $H_2O(l)$.
Calcium	Difficult to ignite; red flame	Moderate reaction; hydroxide formed
Strontium	Difficult to ignite; red flame	Rapid reaction; hydroxide formed
Barium	Difficult to ignite; green flame	Vigorous reaction; hydroxide formed

As you go down Group 2, atomic radius increases and electron shielding increases which leads to increased reactivity with a decrease in first ionisation energy and melting point.

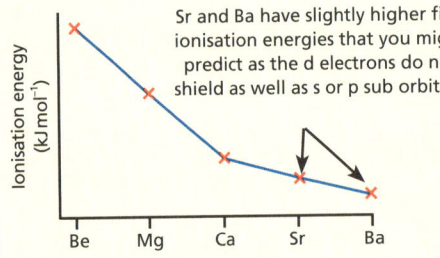

Sr and Ba have slightly higher first ionisation energies that you might predict as the d electrons do not shield as well as s or p sub orbitals.

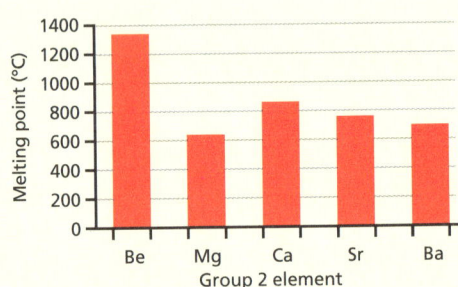

What is the general trend in atomic radius and ionisation energy as you go down Group 2?

Magnesium is an anomaly because it has a different structure to the other metals in the group.

Solubility of Group 2 metal compounds

Solubility is a measure of the mass of a substance that can dissolve in a volume of solvent at a given temperature. The use of the Group 2 metal compounds is linked to solubility.

Hydroxide	Element	Sulfate
Increasing solubility ↓	Mg	↑ Increasing solubility
	Ca	
	Sr	
	Ba	

What are the trends in solubility of Group 2 metal hydroxides and metal sulfates?

Uses of Group 2 metals and their compounds

Medical uses include:

- $BaSO_4$ is used in medical imaging as it is opaque to x-rays.
- $Mg(OH)_2$ and magnesium sulfate, $MgSO_4$, are used as a laxative.

Agricultural uses include:

- $Ca(OH)_2$ is used to increase soil pH.

Industrial uses include:

- Mg in the extraction of titanium from $TiCl_4$, in a displacement reaction. This ensures pure titanium is produced rather than titanium carbide (which would make the material brittle and much less useful).
- CaO, or $CaCO_3$, to neutralise acidic gases in flues.
- Acidified $BaCl_2(aq)$ can be used as an indicator qualitative test for the presence of sulfate ions.

Summary

Group 2

1. Which statement about magnesium is correct? Tick **one** box. [1]

   Magnesium reacts with steam and makes magnesium hydroxide ☐

   Magnesium is a reducing agent in the production of titanium ☐

   Magnesium is part of the p-block in the Periodic Table ☐

   Magnesium sulfate is insoluble in water ☐

2. Which of these properties would you predict radium, Ra, to have? Tick **one** box. [1]

   Forms a highly soluble hydroxide ☐

   Forms a highly soluble sulfate ☐

   Doesn't react with liquid water or steam ☐

   Is a liquid at room temperature ☐

3. This question is about magnesium.

   a) In which block of the Periodic Table is magnesium found? Explain why. [2]

   b) Give an equation, including state symbols, for the reaction of magnesium with steam. [1]

   c) Describe **two** observations when magnesium reacts with steam. [2]

   d) Give **one** use for magnesium metal and **one** use for a magnesium compound. [2]

4. a) Write down the name given to all Group 2 elements. [1]

   b) Describe the trend in melting point moving down Group 2. [1]

   c) Identify the Group 2 element that has an anomalously high melting point. Give a reason for this anomaly. [2]

Group 7

Key words

Key words

Key questions

Where are the halogens found in the Periodic Table?

What is the trend in first ionisation energy and electronegativity as you go down Group 7?

Why is chlorine used in the process of making potable water?

Group 7 (17)

Halogens are very reactive non-metals that form covalent bonds with simple molecular structures with non-metals and ionic lattices with metals.

As you go down Group 7, atomic radius increases, electron shielding increases and electron clouds become more easily polarisable.

So, both the electronegativity and reactivity decrease, but the melting point and boiling point increase.

As you go down the group, the outer-shell electrons are less affected by the nuclear charge and therefore a new electron is less strongly attracted. This causes a decrease in electronegativity, also reducing the oxidising ability and lowering reactivity.

A more reactive halogen can displace a less reactive halogen from its compound, e.g.

$$\text{Chlorine} + \underset{\text{iodide}}{\text{Potassium}} \longrightarrow \underset{\text{Chloride}}{\text{Potassium}} + \text{Iodine}$$

$$Cl_2 + 2KI \longrightarrow KCl + I_2$$

Half equations show more clearly which species is reduced and which is oxidised.

Oxidation half equation: $2I^- \longrightarrow I_2 + 2e^-$

Reduction half equation: $Cl_2 + 2e^- \longrightarrow 2Cl^-$

When a halogen boils, only the weak dipole-dipole interactions are overcome. However, as the molecules get larger, the dipole-dipole forces increase and so does the boiling point.

Uses of chlorine

An aqueous solution of chlorine can react with cold sodium hydroxide. One product, NaClO, is a type of bleach and used to disinfect (kill bacteria).

$$Cl_2(aq) + 2NaOH(aq) \longrightarrow NaCl(aq) + NaClO(aq) + H_2O(l)$$

Chlorine can be used in the production of potable water to ensure it is safe to drink. Chlorine reacts reversibly with water in a disproportionation reaction to make

a powerful oxidising agent that stops pathogens like bacteria from being active.

$$Cl_2(s) + H_2O(l) \rightleftharpoons HCl(aq) + HClO(aq)$$

Society has evaluated the use of chemicals in potable water supplies and concluded that the benefits to health outweigh the toxic effects.

Halides are 1⁻ ions and reducing agents

Halides are identified by using acidified silver nitrate, $AgNO_3$ (aq), where:
- white precipitate that dissolves in dilute ammonia indicates Cl^-
- cream precipitate that dissolves in concentrated ammonia indicates Br^-
- yellow precipitate the doesn't dissolve in ammonia suggests I^-.

Sodium halides and concentrated sulfuric acid

These reactions produce toxic gases and must be completed in a fume cupboard.

Halide ion	Reaction with concentrated sulfuric acid	Observations
Cl^-(aq)	$H_2SO_4(l) + NaCl(s) \longrightarrow HCl(g) + NaHSO_4(s)$	White fumes of HCl gas
Br^-(aq)	$H_2SO_4(l) + NaBr(s) \longrightarrow HBr(g) + NaHSO_4(s)$ $H_2SO_4(l) + 2HBr(s) \longrightarrow Br_2(g) + SO_2(g) + 2H_2O(l)$	Reddish-brown gas of Br_2
I^-(aq)	$H_2SO_4(l) + NaI(s) \longrightarrow HI(g) + NaHSO4(s)$ $2HI(g) + H_2SO_4(l) \longrightarrow I_2(g) + SO_2(g) + 2H_2O(l)$ $6HI(g) + H_2SO_4(l) \longrightarrow 3I_2(g) + S(s) + 4H_2O(l)$ $8HI(g) + H_2SO_4(l) \longrightarrow 4I_2(g) + H_2S(s) + 4H_2O(l)$	Violet/purple vapour of I_2 Yellow solid of S Strong, unpleasant smell of H_2S

Summary

Group 7

(1) Which species is the strongest oxidising agent? Tick **one** box. [1]

F_2 ☐ F^- ☐ I_2 ☐ I^- ☐

(2) Which statement correctly describes the trend down Group 7 from fluorine to iodine? Tick **one** box. [1]

The boiling point of the element decreases ☐

The electronegativity of the element increases ☐

The first ionisation energy of the element decreases ☐

The reducing ability of the element decreases ☐

(3) This question is about chlorine.

a) In which block of the Periodic Table is chlorine found? Explain why. [2]

....

....

b) Why must the amount of chlorine added to drinking water supplies be strictly controlled? [1]

....

....

c) Give a balanced symbol equation for the reaction of chlorine with water. Explain why this is a disproportionation reaction. [3]

....

....

....

....

(4) A student is analysing an unknown solution, X. The student adds acidified silver nitrate to a sample of X and observes a precipitate being formed. The student thinks that solution X contains either chloride ions or bromide ions.

a) Describe a further chemical test using ammonia that the student could use to determine if solution X contains chloride or bromide ions. [3]

....

....

....

b) Describe a further chemical test using chlorine water that the student could use to determine if solution X contains chloride or bromide ions. [3]

....

....

....

....

Period 3

🔑 Key words

❓ Key questions

What are the general trends in melting point for Period 3 oxides?

What is the pH of metal hydroxides?

What is the pH of non-metal hydroxides?

How are acids formed?

Reaction of Period 3 elements with oxygen

General trends in Period 3 oxides:

- Usually metal oxides are basic and non-metal oxides are acidic.
- The melting point of Period 3 oxides generally increase from sodium oxide to silicon dioxide, then decrease from silicon dioxide to chlorine dioxide.

Element	Reaction with oxygen	Structure of oxide	Melting point (°C)	Reaction of oxide with water	Description of solution
Na	$4Na(s) + O_2(g) \longrightarrow 2Na_2O(s)$	Ionic lattice	1132	$Na_2O + H_2O \longrightarrow 2NaOH$	Strong alkali, pH ≈ 13–14
Mg	$2Mg(s) + O_2(g) \longrightarrow 2MgO(s)$	Ionic lattice	2852	$MgO + H_2O \longrightarrow Mg(OH)_2$	Weak alkali, pH ≈ 10
Al	$4Al(s) + 3O_2(g) \longrightarrow 2Al_2O_3(s)$	Ionic lattice with some covalent character	2072	Insoluble	n/a
Si	$Si(s) + O_2(g) \longrightarrow SiO_2(s)$	Giant covalent macromolecule	1710	Insoluble	n/a
P	$P_4(s) + 5O_2(g) \longrightarrow P_4O_{10}(s)$	Simple molecular compound	340	$P_4O_{10} + 6H_2O \longrightarrow 4H_3PO_4$	Weak acid, pH ≈ 1–2
S	$S(s) + O_2(g) \longrightarrow SO_2(g)$	Simple molecular compound	–72	$SO_2 + H_2O \longrightarrow H_2SO_3$	Strong acid, pH ≈ 2–3
S	$2S(s) + 3O_2(g) \longrightarrow 2SO_3(g)$	Simple molecular compound	16.9	$SO_3 + H_2O \longrightarrow H_2SO_4$	Strong acid, pH ≈ 0–1
Ar	No reaction	n/a	n/a	n/a	n/a

Acids

The Brønsted–Lowry theory defines an acid as a proton donor.

Phosphoric(V) acid, H_3PO_4, dissociates in water to release three protons (triprotic acid) and form an anion.	
Sulfuric(IV) acid or sulfurous acid, H_2SO_3, is formed when sulfur dioxide reacts with water. The acid only partially dissociates in water, releasing one proton (monoprotic acid).	
Sulfuric(VI) acid or sulfuric acid, H_2SO_4, is formed when sulfur trioxide reacts with water. This acid dissociates in water, releasing two protons (diprotic acid).	

Acid–base reactions

- Basic oxides react with acids to make a salt and water
- Acidic oxides react with bases to make a salt
- Acidic oxides react with alkalis to make a salt and water
- Amphoteric oxides can react with both acids and bases to make a salt
- Silicon dioxide, SiO_2, cannot dissolve in water but will react with concentrated bases and so is described as an acidic oxide

✔ Summary

Period 3

(1) Which Period 3 oxide would make a solution with the lowest pH value? Tick **one** box. [1]

SiO_2 ☐ SO_2 ☐ P_4O_{10} ☐ SO_3 ☐

(2) Which is the correct formula for the aluminium-containing species formed when aluminium oxide reacts with an excess of water? Tick **one** box. [1]

Al_2O_3 ☐ $Al(OH)_3$ ☐ AlH_3 ☐ $[Al(OH)_4(H_2O)_2]^{1-}$ ☐

(3) This question is about oxides of sulfur.

a) Explain why sulfur trioxide, SO_3, has a higher melting point than sulfur dioxide, SO_2. [3]

b) Write an equation for the reaction between sulfur dioxide and water. [1]

c) Draw the displayed formula for the anion formed when sulfur trioxide reacts with water. [1]

d) Write an equation to show the reaction between magnesium oxide and sulfuric(VI) acid. [1]

(4) Phosphorus and silicon can both react with oxygen to make acidic oxides.

a) Write an equation for the reaction of phosphorus(V) oxide with water. Suggest a pH for the solution formed. [2]

b) Explain why silicon(IV) oxide has a higher melting point than phosphorus(V) oxide. [5]

Transition metals

Key words

Key questions

What are transition metals?

What are the properties of transition metals?

What is a ligand?

What is a complex ion?

Transition metals

Transition metals are elements in the d-block of the Periodic Table with at least one stable ion with an incomplete d sub-shell of electrons.

Transition elements

| | | | Sc | Ti | V | Cr | Mn | Fe | Co | Ni | Cu | Zn | | | | | | |

s-block | d-block elements | p-block

Look at the electronic structures of the Period 3 d-block elements. Sc and Zn are d-block elements but not transition metals:

Sc	$1s^2\ 2s^2\ 2p^6\ 3s^2\ 3p^6\ 4s^2\ 3d^1$
Ti	$1s^2\ 2s^2\ 2p^6\ 3s^2\ 3p^6\ 4s^2\ 3d^2$
V	$1s^2\ 2s^2\ 2p^6\ 3s^2\ 3p^6\ 4s^2\ 3d^3$
Cr	$1s^2\ 2s^2\ 2p^6\ 3s^2\ 3p^6\ 4s^1\ 3d^5$
Mn	$1s^2\ 2s^2\ 2p^6\ 3s^2\ 3p^6\ 4s^2\ 3d^5$
Fe	$1s^2\ 2s^2\ 2p^6\ 3s^2\ 3p^6\ 4s^2\ 3d^6$
Co	$1s^2\ 2s^2\ 2p^6\ 3s^2\ 3p^6\ 4s^2\ 3d^7$
Ni	$1s^2\ 2s^2\ 2p^6\ 3s^2\ 3p^6\ 4s^2\ 3d^8$
Cu	$1s^2\ 2s^2\ 2p^6\ 3s^2\ 3p^6\ 4s^1\ 3d^{10}$
Zn	$1s^2\ 2s^2\ 2p^6\ 3s^2\ 3p^6\ 4s^2\ 3d^{10}$

$\rightarrow$ When forming ions, lose 4s before 3d

| Sc^{3+} [Ar] $4s^0\ 3d^0$ |
| Ti^{3+} [Ar] $4s^0\ 3d^1$ |
| V^{3+} [Ar] $4s^0\ 3d^2$ |
| Cr^{3+} [Ar] $4s^0\ 3d^3$ |
| Mn^{2+} [Ar] $4s^0\ 3d^5$ |
| Fe^{3+} [Ar] $4s^0\ 3d^5$ |
| Co^{2+} [Ar] $4s^0\ 3d^7$ |
| Ni^{2+} [Ar] $4s^0\ 3d^8$ |
| Cu^{2+} [Ar] $4s^0\ 3d^9$ |
| Zn^{2+} [Ar] $4s^0\ 3d^{10}$ |

Transition metals:
- have the usual properties of metals (ductile, malleable, lustrous, conductors)
- form complex ions and coloured ions
- have more than one stable oxidation state, which leads to catalytic activity.

Stable oxidation states of Period 4 d-block elements:

	+1	+2	+3	+4	+5	+6	+7
Sc			Sc^{3+}				
Ti			Ti^{3+}	TiO^{2+}			
V		V^{2+}	V^{3+}	VO^{2+}	VO_2^+		
Cr		Cr^{2+}	Cr^{3+}			$Cr_2O_7^-$	
Mn		Mn^{2+}		MnO_2			MnO_4^-
Fe		Fe^{2+}	Fe^{3+}				
Co		Co^{2+}	Co^{3+}				
Ni		Ni^{2+}					
Cu	Cu^+	Cu^{2+}					
Zn		Zn^{2+}					

Complex ions

Complex ions are:
- a species made from a central metal atom or ion
- surrounded by ligands, which are species that form a co-ordinate bond with a transition metal by donating a pair of electrons.

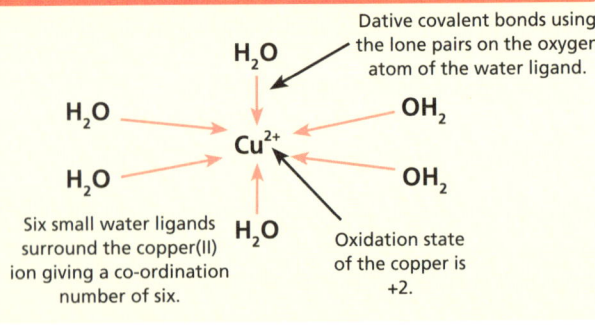

Dative covalent bonds using the lone pairs on the oxygen atom of the water ligand.

Six small water ligands surround the copper(II) ion giving a co-ordination number of six.

Oxidation state of the copper is +2.

Summary

Spec. ref.   3.2.5.1

Transition metals

1 Which is the electron configuration of an atom of a transition metal? Tick **one** box. [1]

[Ar] $3d^1 4s^2$ ☐   [Ar] $3d^{10} 4s^1$ ☐   [Ar] $3d^0 4s^2$ ☐   [Ar] $3d^8 4s^2$ ☐

2 Which of these will not act as a ligand in the formation of a complex ion? Tick **one** box. [1]

Cl^- ☐ H_2O ☐ NH_3 ☐ NH_4^- ☐

3 Copper(II) can form a complex ion with six water ligands.

   a) Explain how the bond is formed between each water molecule and the copper metal. [3]

   b) Deduce the oxidation state of the copper ion and the co-ordination number of the complex ion. [2]

   c) Justify why copper can be classified as both a d-block element and a transition metal. [2]

4 Transition metals have characteristic properties including catalytic action and the formation of complexes with different shapes.

   a) Give the electron configuration of the Sc^{3+} ion. Use your answer to explain why the Sc^{3+} ion is not classified as a transition metal ion. [2]

   b) Explain the meaning of the term 'complex' in relation to transition metals. [2]

Transition metal substitution reactions

Key words

Key questions

What is a substitution reaction?

What is the chelate effect?

How does blood transport oxygen around the body?

Why is carbon monoxide toxic?

Ligands

Formula	Type
H_2O	Monodentate
NH_3	Monodentate
Cl^-	Monodentate
$EDTA^{4-}$	Multidentate

Formula	Type
$H_2NCH_2CH_2NH_2$ Ethane-1,2-diamine	Bidentate
$C_2O_4^{2-}$ Ethanedioate	Bidentate

Substitution reactions

A substitution reaction is when one or more ligands in a complex ion are replaced by different ligands.

The ligands NH_3 and H_2O are similar in size and are uncharged, so it is possible for transition metal ions to undergo a substitution reaction where the co-ordination number is unchanged.

Partial substitution reactions can also occur. The Cl^- ligand is larger than water or ammonia and so the substitution reaction has a change in co-ordination number.

Chelate effect

For substitution reactions to spontaneously occur, $\Delta G \leq 0$ and they are usually entropy-driven when monodentate ligands are substituted for bidentate or multidentate ligands. This reaction increases the entropy as there are more particles and the resulting complex is more thermodynamically stable. For example:

$$[Cu(H_2O)_6]^{2+}(aq) + EDTA^{4-}(aq) \longrightarrow [Cu(EDTA)]^{2-}(aq) + 6H_2O(l)$$

The chelate effect is used:

* in detergents to remove calcium ions from hard water so that they lather well
* to remove heavy metals from polluted water or the body.

Blood

Red blood cells contain a protein called haemoglobin, which is made of haem, an iron(II) complex with a multidentate ligand. A co-ordinate bond can form reversibly with oxygen to make a complex that can then transport oxygen in the blood. However, carbon monoxide is also a ligand and this binds more strongly to haemoglobin than oxygen does. This reduces the oxygen-carrying capacity of the blood and can be fatal.

Summary

Spec. ref. 3.2.5.2

Transition metal substitution reactions

1 A colour change is observed when concentrated hydrochloric acid is added to cobalt(II) chloride solution.

Which type of reaction takes place? Tick **one** box. [1]

Ligand substitution ☐   Redox ☐   Precipitation ☐   Acid–base equilibria ☐

2 Which statement is true about the following reaction? Tick **one** box. [1]

$$[Co(NH_3)_6]^{2+} + 3H_2NCH_2CH_2NH_2) \longrightarrow [Co(H_2NCH_2CH_2NH_2)_3]^{2+} + 6NH_3$$

ΔH is large and positive. ☐

$\Delta G \geqslant 0\,kJ\,mol^{-1}$ ☐

$\Delta S > 0\,J\,mol^{-1}$ ☐

ΔS is unchanged ☐

3 Iron forms many complexes that contain iron in oxidation states +2 and +3. Iron(II) sulfate was dissolved in water.

a) Give the formula of the complex ion made from iron(II) sulfate dissolved in water. [1]

....

b) Iron(III) ions can form a complex with six water molecules and can undergo a ligand substitution reaction with excess concentrated hydrochloric acid.

Write an equation for this reaction. [1]

....

c) Explain why the iron(III) complexes in part b) have two different shapes. [2]

....

....

....

4 A ligand substitution reaction occurs when excess ethane-1,2-diamine is added to a solution containing $[Cu(H_2O)_6]^{2+}$ (aq) ions. Ethane-1,2-diamine is a bidentate ligand and replaces all the water ligands.

a) Explain what is meant by the term 'bidentate ligand'. [1]

....

b) Write an equation for this reaction. [1]

....

c) Explain the thermodynamic reasons why this reaction occurs. [3]

....

....

....

....

Shapes and colours of complex ions

Key words

Shapes of complex ions

Shape	Co-ordination number	Bond angle °	Comments
Octahedral	6	90	Most common shape for complex ions with small ligands
Tetrahedral	4	109.5	Second most common shape for a complex ion usually made with larger ligands (e.g. Cl⁻)
Square planar	4	90	Cyanide ions (CN^-) are the most common ligands found in this shape
Linear	2	180	$[Ag(NH_3)_2]^+$ is in Tollens' reagent

Key questions

What is the most common shape for a complex ion?

Isomerism

Octahedral complexes can show:

- cis–trans isomerism (a special case of E–Z isomerism) with monodentate ligands e.g.

- optical isomerism with bidentate ligands e.g.

What types of isomerism can octahedral complexes show?

$$trans\ [Cr(NH_3)_4Cl_2]^+ \quad cis\ [Cr(NH_3)_4Cl_2]^+$$

Square planar complexes can display cis–trans isomerism, e.g. the anti-cancer drug Cisplatin used in chemotherapy. Only the cis-isomer has a therapeutic use as Cisplatin binds to DNA in cells and prevents them from dividing and causes them to die.

Cisplatin
cis or Z form

Transplatin
trans or E form

Colour and colorimetry

What is Cisplatin?

When a transition metal ion is in isolation, all the d sub-shells are at the same energy level. In a complex ion, the ligand electrons repel some d sub-shells and this causes a split.

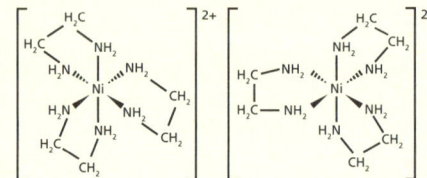
Energy — d orbitals without ligands — ΔE — d orbitals with ligands

When light shines on the solution, some of the wavelengths of visible light are absorbed to promote electrons from the ground state (lower d sub-shell) to an excited state (higher d sub-shell). You can observe the complementary colour of the absorbed wavelengths.

The energy change can be calculated using

Why are solutions of transition metal ions coloured?

$$\Delta E = hv = \frac{hc}{\lambda}$$

where v = frequency of light absorbed (unit s⁻¹ or Hz); h = Planck's constant 6.63×10^{-34} (J s); ΔE = energy difference between split orbitals (J); c = speed of light 3.00×10^8 (m s⁻¹); λ = wavelength of light absorbed (m)

Colour changes are observed when ΔE changes. This can happen if the oxidation state changes, the co-ordination number changes and/or ligands are substituted.

A simple colorimeter can be used to measure the absorption of visible light. The characteristic absorption can be used to identify a transition metal ion and the absorbance of visible light can be used to determine the concentration of the solution.

Summary

1 Which of the following complex ions are square planar? Tick **one** box. [1]

$CuCl_4^{2-}$ ☐ $[CoCl_4]^{2-}$ ☐ $[Pt(NH_3)_2Cl_2]$ ☐ $[Cu(NH_3)_4(H_2O)_2]^{2+}$ ☐

2 A solution absorbs light with wavelengths corresponding to orange.

Which ion is most likely to be in the solution? Tick **one** box. [1]

Fe^{2+} ☐ Fe^{3+} ☐ Cu^{2+} ☐ MnO_4^{2-} ☐

3 Cobalt forms many complexes that contain cobalt in oxidation states +2. Cobalt(II) sulfate is dissolved in water.

a) Give the formula and the shape of the complex ion made from cobalt(II) sulfate dissolved in water. [2]

....

....

b) When concentrated hydrochloric acid is added to the solution, a ligand exchange reaction occurs to form $[CoCl_4]^{2-}$

Explain why the colour of the solution changes from pink to blue. [5]

....

....

....

....

....

....

c) Suggest the bond angle in $[Co(H_2O)_6]^{2+}$ [1]

....

4 Fungicides often contain copper(II) compounds. They can enter the water supply and cause problems because they are toxic in high concentrations.

Samples of water can be tested for copper(II) ions by forming a blue aqueous complex with $EDTA^{4-}$ ions and analysing the sample with a colorimeter.

Outline the practical steps that you would follow, using colorimetry, to determine the concentration of this complex in a sample of water. [5]

....

....

....

....

....

....

....

Oxidation states of transition metals

🔑 Key words

❓ Key questions

What reagents can you use to reduce a transition metal ion?

What affects redox potentials of transition metal ions undergoing reduction?

What is a redox titration?

Why do you not need to use an indicator in many redox titrations?

Oxidation states

Variable oxidation states are seen in transition elements. In general:

- across a period, the relative stability of the +2 state increases compared to the +3 state
- high oxidation states tend to be polyatomic ions and are often oxidising agents
- low oxidation states tend to be monatomic ions and are often reducing agents

Oxidation state	+2	+3	+4	+5
	V^{2+}	V^{3+}	VO^{2+}	VO_2^+
Colour	Purple	Green	Blue	Yellow

Zn can react with concentrated HCl to form a reducing agent. The Zn is oxidised and releases electrons that can reduce a transition metal ion, leading to a colour change.

Redox potential is a measure of how easily a chemical species loses or gains electrons. For a transition metal ion undergoing reduction, it is influenced by:

- pH – changing $[H^+]$ affects the position of equilibrium, e.g. reduction of Mn in MnO_4^- is more likely to occur in acidic solutions.
- ligands – different ligands have different co-ordinate bond strengths and different spatial arrangements; and the chelation effect affects the electron density on the metal ion.

Tollens' reagent

Acidified silver nitrate solution contains $[Ag(NH_3)_2]^+$ that can reduce to form a precipitate of Ag(s) and a carboxylic acid in the presence of an aldehyde. However, with a ketone there is no observable change and so this can be used as a chemical test to distinguish between carbonyl compounds.

$$[Ag(NH_3)_2]^+(aq) + e^- \longrightarrow Ag(s) + 2NH_3(g)$$

Redox titrations

The burette has an oxidising agent (titrant) and this oxidises the substance in the conical flask (analyte). No indicator is needed as the end-point is clearly visible due to the change in colour of the transition metal ion.

Redox titrations using MnO_4^- need to occur under acidic conditions otherwise side reactions occur that affect the volume of oxidising agent added. H_2SO_4 is used as:

- weak acids can't give high enough $[H^+]$
- HCl ionises and Cl^- is oxidised by MnO_4^-
- HNO_3 ionises, and the nitrate ion is an oxidising agent and can react with analytes such as Fe^{2+}.

A 250 cm³ standard solution of sodium ethanedioate is made by dissolving 162 mg of $Na_2C_2O_4$ (M_r = 134.0) in water. 25.00 cm³ of this warmed solution is titrated with acidified potassium manganate(VII) solution and the mean accurate titre is 23.85 cm³. The equation for this reaction is: $2MnO_4^- + 16H^+ + 5C_2O_4^{2-} \longrightarrow 2Mn^{2+} + 8H_2O + 10CO_2$

Use this information to calculate the concentration of the potassium manganate(VII) solution.

Amount of $Na_2C_2O_4 = \frac{0.162}{134.0} = 0.00121$ mol in 250 cm³, so 0.000 121 mol in 25 cm³

Therefore, $\frac{2}{5} \times 0.00121 = 4.84 \times 10^{-5}$ moles of MnO_4^-

Concentration of $MnO_4^- = \frac{4.84 \times 10-5) \times 1000}{23.85} = 0.002\,03$ mol dm⁻³

✔ Summary

Oxidation states of transition metals

1. Which of these compounds can decolourise acidified potassium manganate(VII) solution? Tick **one** box. [1]

   $FeSO_4$ ☐

   $Fe_2(SO_4)^3$ ☐

   $CuSO_4$ ☐

   CH_3COOH ☐

2. In which complex ion does vanadium have an oxidation state of +4? Tick **one** box. [1]

   VO_2^+ ☐

   VO^{2+} ☐

   NH_4VO_3 ☐

   $[V(H_2O)_6]^{2+}$ ☐

3. Tollens' reagent is formed by the addition of aqueous ammonia to aqueous silver nitrate.

   a) Identify the silver-containing species present in the colourless solution. [1]

   b) Give **one** use of this reaction. [1]

4. Hydrogen peroxide, H_2O_2, is an oxidising agent that is used in hair bleach.

   The half-equation for the oxidation of hydrogen peroxide is $H_2O_2 \longrightarrow O_2 + 2H^+ + 2e^-$

   A sample of hair bleach solution is diluted to 5.00% with water.

   A $25.0\,cm^3$ sample of the diluted acidified hair bleach is titrated with $0.0200\,mol\,dm^{-3}$ potassium manganate(VII) solution and the mean accurate titre is $35.75\,cm^3$.

   Calculate the concentration of hydrogen peroxide in the original hair bleach solution. [6]

Transition metals as catalysts

Key words

Key questions

What are catalysts?

Why are transition metal ions good catalysts?

What is the difference between heterogenous and homogeneous catalysts?

What is autocatalysis?

Catalysts

Catalysts provide an alternative reaction pathway with a lower activation energy, so increase the rate of reaction. Transition metals are good catalysts because:
- their incomplete d sub-shell allows e^- to be easily transferred
- they have more than one stable oxidation state.

Heterogenous catalysts

Heterogenous catalysts are in a different phase to the reactants. The catalyst reduces the activation energy by holding the reactant(s) in a favourable geometry and weakening the reactant bonds.

As this is a surface reaction, using an inert support medium maximises the surface area of the catalysts. This is more sustainable as less catalyst is needed.

Examples of heterogeneous catalysts:
- Vanadium(V) oxide, V_2O_5, is used in the Contact Process to make sulfuric acid. The vanadium compound allows an alternative two-step process to occur:

$$SO_2 + V_2O_5 \longrightarrow SO_3 + V_2O_4 \quad \text{then} \quad 2V_2O_4 + O_2 \longrightarrow 2V_2O_5$$

- Iron, Fe, is used in the Haber Process to make ammonia.

The active sites can become blocked by impurities in the reactant mixture and products that do not effectively desorb. This reduces efficiency and increases costs.

Homogeneous catalysts

A homogeneous catalyst is in the same phase as the reactants and the reaction happens via an intermediate species.

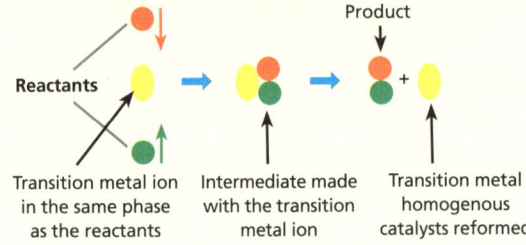

Reactants

Product

Transition metal ion in the same phase as the reactants

Intermediate made with the transition metal ion

Transition metal homogenous catalysts reformed

Explain how iron(II) can increase the rate of reaction between persulfate and iodide ions.

The rate of reaction between persulfate and iodide ions is slow because a collision needs to occur between two negative ions. The equation is: $S_2O_8^{2-} + 2I^- \longrightarrow 2SO_4^{2-} + I_2$

Fe^{2+} can catalyse the reaction because it is a homogenous catalyst with an electrode potential between the electrode potentials of the two reactants and it provides an alternative reaction pathway with a lower activation energy:

Stage 1: $S_2O_8^{2-} + Fe^{2+} \longrightarrow 2SO_4^{2-} + 2Fe^{3+}$ Stage 2: $2I^- + 2Fe^{3+} \longrightarrow 2Fe^{2+} + I_2$

Autocatalysis

Autocatalysis is a positive feedback loop that is created when one of the products of the reaction can act as a catalyst.

Manganate (VII) ions and ethanoate ions react.

$$2MnO_4^- + 16H^+ + 5C_2O_4^{2-} \longrightarrow 2Mn^{2+} + 10CO_2 + 8H_2O$$

Explain how the rate of reaction changes as this reaction proceeds.

Initially, the reaction is slow as a collision has to occur between two negative species. However, Mn^{2+} is a homogenous catalyst and is one of the products of this reaction. Therefore, as more of the reaction occurs, there is an increase in the concentration of the catalyst and an increase in the rate of reaction until all the reactant is used up. This is an example of autocatalysis.

✓ Summary

RETRIEVE

Transition metals as catalysts

A-level only

1) Which transition metal species is used as a catalyst in the Contact Process? Tick **one** box. [1]

V_2O_5 ☐ Fe ☐ Ni ☐ MnO_4^- ☐

2) Which reaction pair shows autocatalysis? Tick **one** box. [1]

Manganate(VII) ions and ethanoate ions ☐

Persulfate and iodide ions ☐

Hydrogenation of ethene ☐

The Haber Process ☐

3) The Haber Process is an industrial process used to make ammonia.

a) Name the catalyst used in the Haber Process. .... [1]

b) The catalyst used in the Haber Process is a heterogeneous catalyst.

Give the meaning of the term 'heterogeneous catalyst'. [4]

4) Acidified potassium manganate(VII) undergoes an autocatalysed redox reaction with sodium ethanedioate.

$2MnO_4^- + 16H^+ + 5C_2O_4^{2-} \longrightarrow 2Mn^{2+} + 10CO_2 + 8H_2O$

a) Identify the species which acts as a catalyst. .... [1]

b) Explain how this species can act as a catalyst in this reaction. [4]

Reactions of ions in aqueous solution

Key words

Key questions

What does 'amphoteric' mean?

What is a precipitate?

Why are transition metal ions good catalysts?

Why are 3$^+$ transition metal ions more acidic than 2$^+$ metal ions?

Why do Fe^{2+} precipitates change colour on standing in air?

Acid–base reactions of aqueous transition metal ions

$[M(H_2O)_6]^{3+}$ is more acidic than $[M(H_2O)_6]^{2+}$ because 3$^+$ metal ions have a greater charge-to-size ratio and have greater polarising power. This means that it attracts the water molecule more strongly and weakens the O–H bond, so H$^+$ ions are more easily released.

Some metal ions, such as $Al(H_2O)_3(OH)_3(s)$, are amphoteric (they can react as both an acid and a base):

- As an acid: $Al(H_2O)_3(OH)_3(s) + OH^-(aq) \longrightarrow [Al(OH)_4]^-(aq) + 3H_2O(l)$
- As a base: $Al(H_2O)_3(OH)_3(s) + 3H^+(aq) \longrightarrow [Al(H_2O)_6]^{3+}(aq)$

Reactions of aqueous transition metal ions

Reagent	H$_2$O(l) (aq)	NaOH(aq) OH$^-$ (aq)	NH$_3$(aq) NH$_3$	Na$_2$CO$_3$(aq) CO$_3$$^{2-}$
Fe^{2+}	$[Fe(H_2O)_6]^{2+}$	$[Fe(H_2O)_6]^{2+}(aq) + 2OH^-(aq) \longrightarrow$ $Fe(H_2O)_4(OH)_2(s) + 2H_2O(l)$ **Green precipitate** that goes **brown** on standing in air and no further change on excess NaOH.	$[Fe(H_2O)_6]^{2+}(aq) + 2NH_3(aq) \longrightarrow$ $Fe(H_2O)_4(OH)_2(s) + 2NH_4^+(l)$ **Green precipitate** that goes **brown** on standing in air and no further change on excess NH$_3$.	$[Fe(H_2O)_6]^{2+}(aq) + CO_3^{2-}(aq) \longrightarrow$ $FeCO_3(s) + 6H_2O(l)$ **Green precipitate**
Fe^{3+}	$[Fe(H_2O)_6]^{3+}$ In solution, Fe(III) undergoes hydrolysis and appears **yellow/ brown**.	$[Fe(H_2O)_6]^{3+}(aq) + 3OH^-(aq) \longrightarrow$ $Fe(H_2O)_3(OH_3)(s) + 3H_2O(l)$ **Brown precipitate** with no further change on excess NaOH.	$[Fe(H_2O)_6]^{3+}(aq) + 3NH_3(aq) \longrightarrow$ $Fe(H_2O)_3(OH_3)(s) + 3NH_4^+ (l)$ **Brown precipitate** with no further change on excess NH3.	$2[Fe(H_2O)_6]^{3+}(aq) + 3CO_3^{2-}(aq) \longrightarrow$ $2Fe(OH)_3(H_2O)_3(s) + 3CO_2 + 3H_2O(l)$ **Brown precipitate** and effervescence.
Cu^{2+}	$[Cu(H_2O)_6]^{2+}$	$[Cu(H_2O)_6]^{2+}(aq) + 2OH^- (aq) \longrightarrow$ $Cu(H_2O)_4(OH)_2(s) + 2H_2O(l)$ **Blue precipitate** with no further change on excess NaOH.	$[Cu(H_2O)_6]^{2+}(aq) + 2OH^- (aq) \longrightarrow$ $Cu(H_2O)_4(OH)_2(s) + 2H_2O(l)$ **Blue precipitate** that re-dissolves to form a **deep blue solution** with excess NH$_3$. $Cu(OH)_2(H_2O)_4(s) + 4NH_3(aq) \longrightarrow$ $[Cu(NH_3)_4(H_2O)_2]^{2+}(aq) + 2H_2O(l) + 2OH^-(aq)$	$[Cu(H_2O)_6]^{2+}(aq) + CO_3^{2-}(aq) \longrightarrow$ $CuCO_3(s) + 6H_2O(l)$ **Blue-green precipitate**
Al^{3+}	$[Al(H_2O)_6]^{3+}$	$[Al(H_2O)_6]^{2+}(aq) + 3OH^-(aq) \longrightarrow$ $Al(H_2O)_3(OH)_3(s) + 3H_2O(l)$ **White precipitate** that re-dissolves into colourless solution with excess OH$^-$.	$[Al(H_2O)_6]^{2+}(aq) + 3OH^-(aq) \longrightarrow$ $Al(H_2O)_3(OH)_3(s) + 3H_2O(l)$ **White precipitate** and no further change with excess NH$_3$.	$2[Al(H_2O)_6]^{3+}(aq) + CO_3^{2-}(aq) \longrightarrow$ $2Al(OH)_3(H_2O)_3(s) + 3CO_2 + 3H_2O(l)$ **White precipitate** and effervescence.

✔ Summary

Reactions of ions in aqueous solution

1 Which type of precipitate forms when a solution of sodium carbonate is added to a solution of aluminium(III) nitrate?

Tick **one** box. [1]

A white precipitate of aluminium(III) carbonate ☐

A white precipitate of aluminium(III) hydroxide ☐

A white precipitate of aluminium(III) carbonate and bubbles of carbon dioxide gas ☐

A coloured precipitate of aluminium(III) hydroxide and bubbles of carbon dioxide gas ☐

2 Which compound gives a deep blue solution when an excess of dilute aqueous ammonia is added? Tick **one** box. [1]

$CuCl_2$ ☐ $FeCl_3$ ☐ $AlCl_3$ ☐ $FeCl_2$ ☐

3 A solution, X, contains $[Fe(H_2O)_6]^{2+}$ ions.

a) Give the colour of solution X. .... [1]

b) Solution X reacts with aqueous ammonia.

Write an equation for this reaction and describe what you would observe. [3]

....

....

....

c) Explain what happens to the product of the reaction described in part b) when it is left to stand in air. [2]

....

....

....

4 Copper(II) chloride can form a blue solution containing $[Cu(H_2O)_6]^{2+}$ ions. This solution can react with different substances.

$[Cu(H_2O)_6]^{2+}(aq)$ —**Reaction 1**→ Pale blue precipitate —**Reaction 2**→ Deep blue solution

↓ **Reaction 3**

Green-blue precipitate

For each reaction, identify the reagent and write an equation for the reaction. [8]

....

....

....

....

....

....

....

....

Introduction to organic chemistry

Key words

Key words

Representing organic compounds

Organic chemistry is the study of carbon-containing compounds. Organic compounds can be represented in different ways.

Structural formula	Molecular formula	Empirical formula
Represents the arrangement of atoms in a molecule. CH_3CH_3	The number of each type of atom in a compound. C_2H_6	Smallest whole number ratio of atoms in a compound. CH_3
Displayed formula	**Skeletal formula**	**General formula**
Every atom and every bond in the molecule.	Doesn't show H and represents C at the corners and ends of lines.	Represents the composition of a homologous series of compounds. C_nH_{2n+2}

Key questions

What is a homologous series?

What is a functional group?

What is the root in the name of an organic compound?

What information does the suffix in the name of an organic compound give?

What information does the prefix in the name of an organic compound give?

A **homologous series** is a family of chemicals that have the same functional group, general formula and successive members differ by $-CH_2$.

Naming organic compounds

1. Find the longest carbon chain and determine the root:

Number of carbon atoms	1	2	3	4	5	6
Root	meth	eth	prop	but	pent	hex

2. Identify the functional group to determine the suffix. For alkenes, alcohols and ketones, a number will be included before the prefix to show which carbon atom(s) are attached to the functional group.

Functional group	Saturated hydrocarbon	C=C	–OH	ROH	ROR'	–COOH
Homologous series	alkane	alkene	alcohol	aldehyde	ketone	carboxylic acid
Suffix	-ane	-ene	-ol	-al	-one	-oic acid

Halogenoalkanes have halogens as functional groups and do not change the suffix of the root of the name from -ane. They have a prefix fluoro-, chloro-, bromo- or iodo- depending on the halogen present and include a number to show which carbon atom they are found on.

Add any prefixes, including a number to show which carbon the side chain is found:

Number of hydrocarbon atoms in the side chain	2	1	3	4
Prefix	ethyl	methyl	propyl	butyl

If there are multiple side groups, use the prefix di- (2), tri- (3) or tetra- (4). Separate numbers by a comma and put a hyphen between numbers and letters.

The structural formula of a hydrocarbon is:
$CH_3CH(CH_3)CH(C_2H_5)CH_2CH_2CH_3$

The name of this compound is 3-ethyl-2-methylhexane, not 2-methyl-3-ethylhexane, as there is more than one type of side group and they should be listed in alphabetical order.

If the structure is a ring, then cyclo- is added to the start of the name.

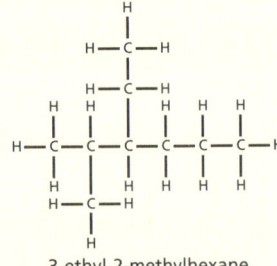

3-ethyl-2-methylhexane

Summary

Introduction to organic chemistry

1 An organic substance has this skeletal formula:

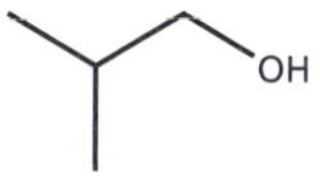

What is the name of this substance? Tick **one** box. [1]

2-methylpropan-1-ol ☐   2-methylethan-1-ol ☐

Butanol ☐   Methylethanol ☐

2 The diagram shows butan-1-ol.

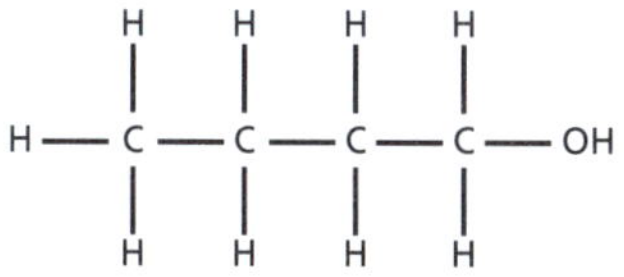

Which type of formula is shown? Tick **one** box. [1]

Displayed ☐   Structural ☐   Molecular ☐   Empirical ☐

3 Crude oil is a rich source of hydrocarbons that can be used in fuels. One branched alkane found in crude oil is 3-ethyl-4-methylhexane, which can be used in petrol.

a) Give the general formula of an alkane. [1]

b) Draw the skeletal formula of 3-ethyl-4-methylhexane. [1]

c) Determine the empirical formula of 3-ethyl-4-methylhexane. [1]

4 This question is about hex-2-ene.

a) Give the molecular formula of hex-2-ene. [1]

b) Identify the functional group in hex-2-ene. [1]

c) Hex-2-ene can undergo an addition reaction with chlorine to make a saturated halogenoalkane.

i) Name the product of this reaction. [1]

ii) Draw its displayed formula. [1]

Reaction mechanisms

Key words

Key questions

What is a reaction mechanism?

What does a single-headed arrow represent in a reaction mechanism?

What does a double-headed arrow represent in a reaction mechanism?

What is a free radical?

Introduction to reaction mechanisms

Scientific models, which are simplified versions of what is really happening, can help you to understand observations and make predictions. Reaction mechanisms are a scientific model that can be used to represent reactions of organic compounds.

In an organic reaction, there are usually multiple steps that occur in a specific order. The reaction mechanism focuses on the movement of electrons in each step and shows the changes to the structure of the organic species in each step.

In a reaction mechanism:
- a single dot represents an unpaired electron, e.g. Cl$^\bullet$ is a chlorine radical with an unpaired electron.
- two dots represent an electron pair, e.g. a water molecule with two lone pairs of electrons
- curly arrows represent the movement of electrons.

Covalent bonds

Covalent bonds can be broken in a reaction mechanism and this is shown by a curly arrow that starts from the bond. There are two types of bond-breaking process:
- Homolytic fission, which is represented by a single-headed, curly arrow that shows the movement of one electron, e.g. breaking the covalent bond in a chlorine molecule to form radicals:

$$Cl - Cl \longrightarrow 2Cl\bullet$$

- Heterolytic fission, which is represented by a double-headed arrow that shows the movement of two electrons, e.g. breaking the covalent bond in a hydrogen chloride molecule to form ions:

movement of electron pair represented by a double-headed arrow

$$H - Cl \longrightarrow H^+ + Cl^-$$

The formation of covalent bonds can be represented by two single headed curly arrows that each starts from a lone electron on the two different radicals. This occurs in the termination step of a free radical mechanism (see below), e.g.

$$Cl\bullet \quad \bullet CH_3 \longrightarrow Cl - CH_3$$

Free radical mechanisms

Alkanes can react with halogens through a free radical mechanism. This mechanism has three parts:

1. Initiation (making the radical via homolytic fission):

free radical

$$Br - Br \xrightarrow{h\nu} 2 \, Br\bullet$$

2. Propagation (different radicals being made):

$$H - \overset{\overset{\displaystyle H}{|}}{\underset{\underset{\displaystyle H}{|}}{C}} - H \quad \bullet Br \longrightarrow H - \overset{\overset{\displaystyle H}{|}}{\underset{\underset{\displaystyle H}{|}}{C}}\bullet + H - Br$$

3. Termination (two radicals coming together and forming a covalent bond):

$$H_3C\bullet \quad \bullet Br \longrightarrow H_3C - Br$$

Summary

Reaction mechanisms

1 Which equation shows the initiation step in a reaction mechanism? Tick **one** box. [1]

$Cl_2 \rightarrow 2Cl^{\bullet}$ ☐

$Cl_2 + 2e^- \rightarrow 2Cl^-$ ☐

$2Cl^{\bullet} \rightarrow Cl_2$ ☐

$CH_3^{\bullet} + Cl^{\bullet} \rightarrow CH_3Cl$ ☐

2 What is shown by a double-headed, curly arrow in a reaction mechanism? Tick **one** box. [1]

A free radical ☐

The movement of two electrons ☐

The movement of one electron ☐

The formation of a lone pair of electrons from two free radicals joining together ☐

3 Trifluoromethane reacts with bromine to form bromotrifluoromethane, which is used as an aviation fire extinguisher.

a) What type of reaction forms bromotrifluoromethane? [2]

....

b) State **one** condition necessary for the reaction to begin. [1]

....

c) Use curly arrows to show the first step in the mechanism for this reaction. [2]

4 The figure below shows an incomplete mechanism for the dehydration of ethanol.

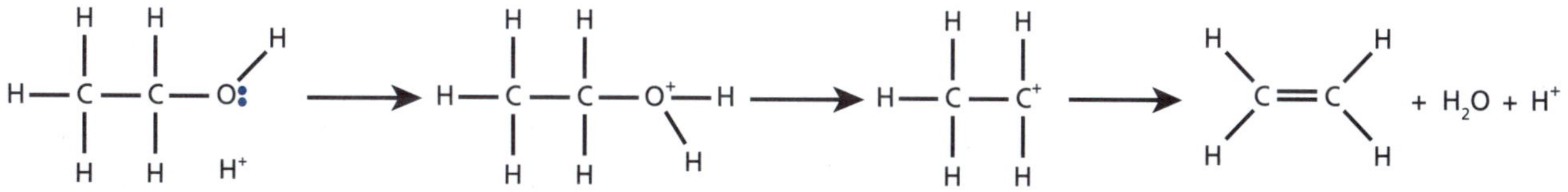

Complete the mechanism in the figure above by adding three curly arrows and any relevant lone pairs of electrons. [3]

Isomerism

Structural isomerism

Structural isomers have the same molecular formula but a different arrangement of atoms.

Chain isomers	Position isomers	Functional group isomers
Their longest carbon chains are different. **Example** 3,3-dimethylpentane 2-methylhexane	They have the same functional group on a different part of the longest hydrocarbon chain. **Example** 1-bromopropane 2-bromopropane	They have different functional groups. **Example** propanal propanone

Stereoisomerism

Stereoisomers have the same molecular and structural formulae but a different arrangement of atoms in space.

In 1,2-dichloroethene, the C=C is a planar bond that doesn't allow rotation and this leads to E–Z isomerism.

Z-1,2-dichloroethene   E-1,2-dichloroethene

Cahn–Ingold–Prelog (CIP) priority rules are used to name the isomers:
- Focus on the C=C and the groups that come off each carbon atom to assess their priority.
- The highest atomic number gets priority 1 (i.e. Cl in the case above) and the second highest gets priority 2.
- If the two priority 1 groups are on the same side of the molecule, it is the Z isomer (zusammen, German for 'together'). If they are on opposite sides, it is the E (entgegen, German for 'opposite') isomer.

Summary

Isomerism

(1) Which alkene shows E–Z isomerism? Tick **one** box. [1]

4-methylpent-2-ene ☐

Ethene ☐

2-methylpropene ☐

2,3-dimethylbut-2-ene ☐

(2) How many structural isomers are there with the molecular formula C_5H_{12}? Tick **one** box. [1]

1 ☐　　　　2 ☐　　　　3 ☐　　　　4 ☐

(3) Propan-2-ol has the molecular formula C_3H_8O.

a) Draw the displayed formula of a position isomer of propan-2-ol. [1]

b) When heated with acidified potassium dichromate(VI), propan-2-ol is oxidised to form propanone.

Name a functional group isomer of propanone. [1]

c) Explain why propan-2-ol **cannot** show E-Z isomerism. [2]

(4) An organic substance has the molecular formula $C_5H_{10}O$. The diagrams show stereoisomers of $C_5H_{10}O$.

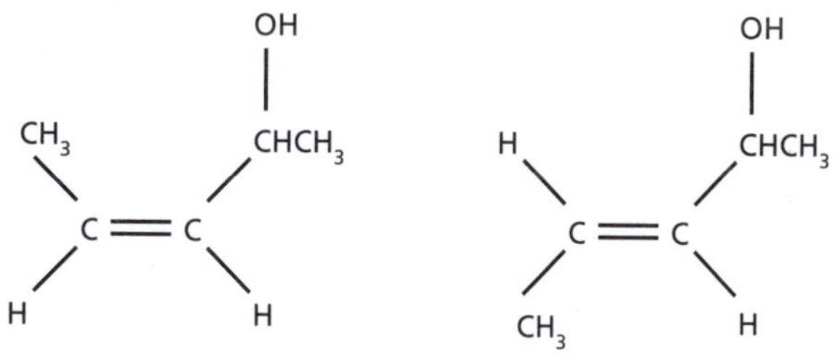

Explain how to determine which isomer is Z and which is E. [5]

Useful hydrocarbons from crude oil

🔑 Key words

❓ Key questions

What are fractions?

What are alkanes?

What is the purpose of cracking?

What is the difference between thermal and catalytic cracking?

Crude oil is a mixture of mainly alkanes

The alkanes in crude oil can be separated using fractional distillation. The alkane homologous series contains saturated hydrocarbons with the general formula C_nH_{2n+2}.

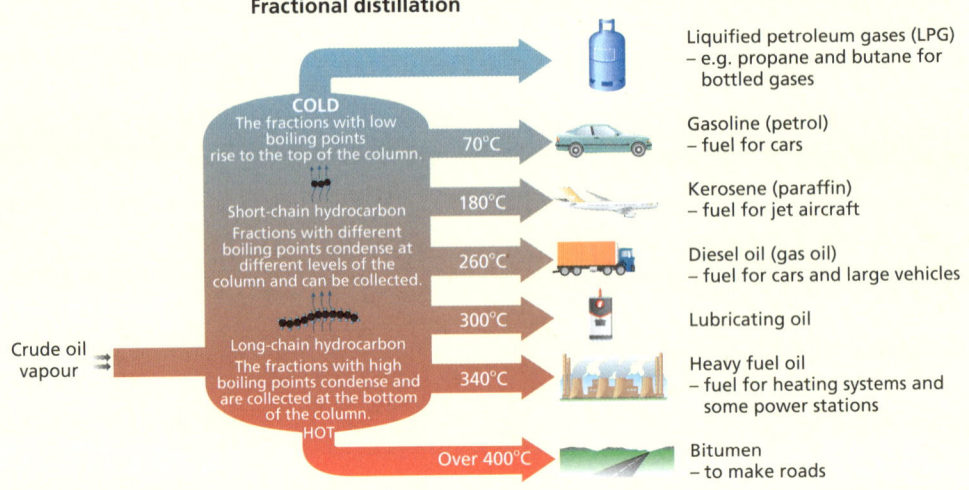

Fractional distillation

COLD
The fractions with low boiling points rise to the top of the column.

Short-chain hydrocarbon
Fractions with different boiling points condense at different levels of the column and can be collected.

Crude oil vapour

Long-chain hydrocarbon
The fractions with high boiling points condense and are collected at the bottom of the column.
HOT

- 70°C — Gasoline (petrol) – fuel for cars
- 180°C — Kerosene (paraffin) – fuel for jet aircraft
- 260°C — Diesel oil (gas oil) – fuel for cars and large vehicles
- 300°C — Lubricating oil
- 340°C — Heavy fuel oil – fuel for heating systems and some power stations
- Over 400°C — Bitumen – to make roads

Liquified petroleum gases (LPG) – e.g. propane and butane for bottled gases

Cracking

Cracking can be used to break down the surplus longer chain hydrocarbons into shorter, more useful alkanes. It can also be used to make alkenes. Cracking is a decomposition reaction that breaks the C–C bonds in the long-chain alkanes.

Thermal cracking makes a lot of alkenes but requires high temperature and pressure, and must be done quickly to reduce the amount of carbon and hydrogen produced. A simplified version of the free radical mechanism is shown where R and R' are ways of showing a hydrocarbon chain.

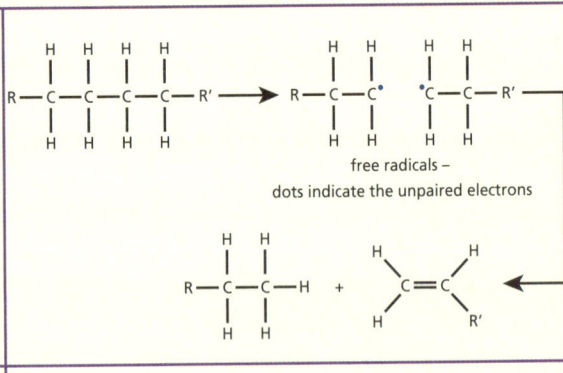

free radicals – dots indicate the unpaired electrons

Catalytic cracking makes a lot of short-chain alkanes for use as motor fuels. It also produces aromatic hydrocarbons. Catalytic cracking is cheaper than thermal cracking but still requires high temperature with a zeolite catalyst and a slightly raised pressure. The reaction can be completed in the lab.

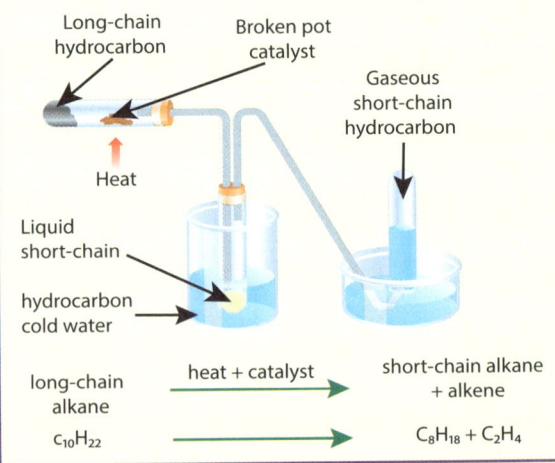

Long-chain hydrocarbon

Broken pot catalyst

Gaseous short-chain hydrocarbon

Heat

Liquid short-chain hydrocarbon

cold water

long-chain alkane $\xrightarrow{\text{heat + catalyst}}$ short-chain alkane + alkene

$C_{10}H_{22} \longrightarrow C_8H_{18} + C_2H_4$

Useful hydrocarbons from crude oil

(1) Which is a fraction of crude oil? Tick **one** box. [1]

A pure hydrocarbon with an exact boiling point ☐

A mixture of hydrocarbons with similar boiling points ☐

A mixture of structural isomers of alkanes ☐

Short-chain hydrocarbons ☐

(2) Which statement is **not** true about alkanes? Tick **one** box. [1]

Alkanes are hydrocarbons ☐

Alkanes are saturated ☐

Alkanes contain the C=C functional group ☐

Petroleum is a mixture consisting mainly of alkanes ☐

(3) Poly(ethene) is an addition polymer. The three key steps to making poly(ethene) from crude oil are shown.

$$\text{crude oil} \xrightarrow{\text{step 1}} \text{naphtha} \xrightarrow{\text{step 2}} \text{ethene} \xrightarrow{\text{step 3}} \text{poly(ethene)}$$

a) Identify the process used in step 1. .... [1]

b) Explain the purpose of step 2. [2]

c) Explain why naphtha boils over a range of temperatures. [2]

(4) Saturated hydrocarbons, such as hexadecane, $C_{16}H_{34}$, can be used to make fuels for central heating systems.

a) Give the meaning of the term 'saturated'. [1]

b) i) Name the substance from which hexdecane is obtained. [1]

ii) State the name of the process used to obtain hexadecane from this substance. [1]

c) The hydrocarbon $C_{16}H_{34}$ can undergo catalytic cracking to form octane, butane and one alkene.

i) Write a balanced symbol equation for this reaction. [1]

ii) Give the reaction conditions. [2]

d) Explain why oil companies need to crack heavy fractions. [2]

Reactions of alkanes

ORGANISE

Key words

Key questions

What is the difference between complete and incomplete combustion?

What are the environmental problems of using alkanes as fuels?

What condition is needed for the chlorination of alkanes?

What mechanism is used in the chlorination of methane?

Combustion is an exothermic oxidation reaction

There are two types of combustion:

- **Complete combustion.** Excess oxygen, fuel is fully oxidised and maximum energy released.

$$C_xH_y + (x\tfrac{y}{4})\, O_2 \rightarrow xCO_2 + \tfrac{y}{2}H_2O$$

- **Incomplete combustion.** Limited oxygen, partial oxidation and limited energy released. Incomplete combustion of hydrocarbons produces H_2O and a mixture of carbon containing products (CO, C and CO_2).

Pollution

When alkanes are used as fuels, they are combusted. The products of the combustion can lead to environmental problems:

- Carbon dioxide – a greenhouse gas and the main contributor to global climate change.
- Carbon (soot) – a particulate made from the incomplete combustion of alkanes that contributes to global dimming.
- Sulfur dioxide – a gas produced from the oxidation of sulfur impurities in the fuels and can lead to acid rain. This can be removed from flue gases by reacting the acidic gases with a base such as CaO or $CaCO_3$.
- Nitric oxide (NO) and nitrogen dioxide (NO_2) – collectively known as NO_x, these gases are produced in car engines as the heat and pressure causes the oxidation of nitrogen in the air, which can lead to acid rain.

Internal combustion engines

Car engines operate at high temperature and pressure. They can form a mixture of pollutants including NO_x, carbon monoxide, carbon and unburned hydrocarbons. Carbon monoxide emissions can be lethal. These gaseous pollutants from internal combustion engines can be removed using catalytic converters.

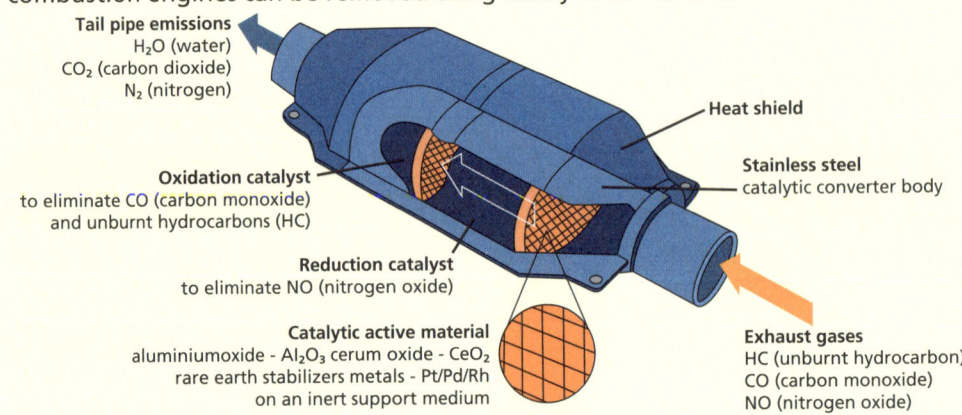

Tail pipe emissions
H_2O (water)
CO_2 (carbon dioxide)
N_2 (nitrogen)

Heat shield

Oxidation catalyst
to eliminate CO (carbon monoxide)
and unburnt hydrocarbons (HC)

Stainless steel
catalytic converter body

Reduction catalyst
to eliminate NO (nitrogen oxide)

Catalytic active material
aluminiumoxide - Al_2O_3 cerum oxide - CeO_2
rare earth stabilizers metals - Pt/Pd/Rh
on an inert support medium

Exhaust gases
HC (unburnt hydrocarbon)
CO (carbon monoxide)
NO (nitrogen oxide)

Chlorination of methane

Chlorine reacts with methane in the presence of ultraviolet light to make a mixture of products. This is a chain reaction involving free radical substitution.

Initiation: $Cl_2 \rightarrow 2Cl^{\bullet}$

Main propagation steps:
$CH_4 + Cl^{\bullet} \rightarrow CH_3^{\bullet} + HCl$
$CH_3^{\bullet} + Cl_2 \rightarrow CH_3Cl + Cl^{\bullet}$

Main termination steps:
$2Cl^{\bullet} \rightarrow Cl_2$
$2CH_3^{\bullet} \rightarrow C_2H_6$
$CH_3^{\bullet} + Cl^{\bullet} \rightarrow CH_3Cl$

e.g. chloromethane, dichloromethane, trichloromethane and tetrachloromethane.

✔ Summary

Reactions of alkanes

(1) Which statement is **not** correct about the pollutant sulfur dioxide? Tick **one** box. [1]

Sulfur dioxide is an acidic gas that makes sulfuric(IV) acid with rainwater ☐

Sulfur dioxide can be removed from car exhausts using a catalytic converter ☐

Sulfur dioxide can be removed from flue gases by reacting with calcium oxide ☐

Sulfur dioxide can be removed from flue gases by reacting with calcium carbonate ☐

(2) Methane reacts with chlorine to form chloromethane.

What is the name of the mechanism for this reaction? Tick **one** box. [1]

Free radical substitution ☐

Free radical addition ☐

Condensation reaction ☐

Elimination ☐

(3) Chloromethane is an important industrial solvent and is made from alkanes.

   a) Chlorine reacts with methane in the presence of which type of radiation? [1]

   b) Explain why a mixture of products is formed when methane reacts with chlorine. [2]

   c) Give the equation for a termination step to produce a molecule of chloromethane. [1]

   d) Suggest a process that could be used to separate the products of the reaction between methane and chlorine. [1]

(4) Petrol is used in car engines and contains a mixture of alkanes, including 3-ethyl-4-methylhexane.

   a) Write an equation for the complete combustion of 3-ethy-4-methylhexane. [2]

   b) Nitrogen oxides (NO_x) are formed when petrol is burned in the internal combustion engines of cars.

      i) State **one** environmental problem that NO_x causes. [1]

      ii) Explain how NO_x can be removed from exhaust gases. [3]

Nucleophiles and halogenoalkanes

Key words

Key questions

What is a nucleophile?

Why are halogenoalkanes susceptive to nucleophilic attack?

What is the functional group in nitrile compounds?

What mechanism is used when a Lewis acid reacts with a halogenoalkane?

Nucleophiles

Nucleophiles can be abbreviated to Nu: and can exist without charge (e.g. NH_3) or as an anion (e.g. OH^-, CN^-). They are chemical species that can form a bond by donating a pair of electrons and a Lewis base.

Halogenoalkanes

The C–X bond is polar as the halogen, X, has a higher electronegativity value than C. There is a slight positive charge, δ^+, which makes it attractive to a nucleophile.

The rate of nucleophilic substitution depends on the C–X bond enthalpy. Strong bonds (C–F) are harder to break and therefore do not react at all, whereas weaker bonds (C–I) cause iodoalkanes to react the fastest.

Nucleophilic substitution reaction to form a nitrile (–C≡N)

Reflux a halogenoalkane with ethanoic potassium cyanide, KCN, for a reaction between an CN^- nucleophile to make a nitrile.

This reaction increases the length of the carbon chain

Nucleophilic substitution reaction to form a primary amine (R–NH₂)

Heat halogenoalkanes with ammonia in a sealed test tube for a two-step reaction between an NH_3 nucleophile to make an amine.

The reaction forms ammonium bromide salt as well as the organic product ethylamine

Nucleophilic substitution reaction to form an alcohol (–OH)

Reflux a halogenoalkane with a dilute aqueous base (e.g. NaOH or KOH) for a reaction between an OH^- nucleophile to make an alcohol. The water present in the solvent causes a hydrolysis reaction but a mixture of products is formed as some elimination reactions also occur.

Primary halogenoalkanes favour substitution reactions

Nucleophilic elimination reaction to form an alkene (C=C)

Heat a halogenoalkane with an ethanoic concentrated base (e.g. NaOH or KOH) for an elimination reaction to form an alkene.

Tertiary halogenoalkanes favour elimination reactions

Summary

Nucleophiles and halogenoalkanes

1 What type of reaction happens between a primary iodoalkane under reflux with an aqueous solution of sodium hydroxide?
Tick **one** box. [1]

Nucleophilic elimination ☐   Free radical substitution ☐

Nucleophilic substitution ☐   Free radical termination ☐

2 Which of the following is **not** a nucleophile? Tick **one** box. [1]

NH_3 ☐ CN^- ☐

OH^- ☐ Cl_2 ☐

3 Sodium hydroxide can react with 1-bromobutane to make different products.

a) Explain how sodium hydroxide produces a nucleophile. [3]

b) Outline a mechanism for the reaction of 1-bromobutane with a dilute aqueous solution of sodium hydroxide.
Give the displayed formula of the organic product. [3]

c) Name the product formed when concentrated ethanoic sodium hydroxide solution is heated and reacts with
1-bromobutane. [1]

4 2-bromo-2-methylpentane is heated with potassium hydroxide dissolved in ethanol. A mixture of alkenes is formed.

a) Deduce the molecular formula of the alkenes formed. [1]

b) Draw the **two** mechanisms to show the formation of the two organic products. Name the two organic products. [6]

c) Explain why halogenoalkanes are susceptible to nucleophilic attack. [2]

Ozone

🔒 Key words

❓ Key questions

What is ozone?

Why is the ozone layer important to humans?

How do CFCs affect the ozone layer?

What are the alternatives to CFCs?

Ozone

Ozone absorbs UVc radiation in the upper atmosphere. This electromagnetic radiation can cause health problems such as cancer, eye damage and suppress the immune system, so its absorption in the upper atmosphere helps to protect humans.

Ozone is constantly forming and decomposing in a natural equilibrium in the stratosphere: $3O_2 \rightleftharpoons 2O_3$

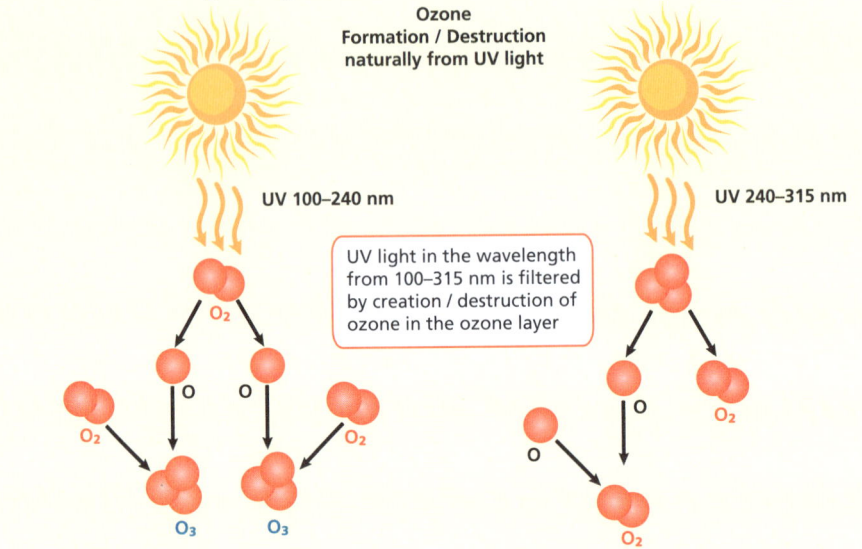

Ozone Formation / Destruction naturally from UV light

UV 100–240 nm

UV 240–315 nm

UV light in the wavelength from 100–315 nm is filtered by creation / destruction of ozone in the ozone layer

- UV light below 240 nm will disturb the bond of the oxygen molecule and form 2 oxygen atoms
- These oxygen atoms will quickly attach to diatomic oxygen molecules to form ozone
- Peak ozone generation occurs at 185 nm wavelength of UV light

- UV light in the 240–316 nm range will disrupt the bond of the ozone molecule and convert this ozone back to diatomic oxygen
- Peak ozone destruction occurs at 254 nm wavelenght of UV light

Ozone depletion

Halogenoalkanes, including chlorofluorocarbons (CFCs), have been used since the 19th century for applications including solvents, refrigerants and propellants. Once used and released, they mix with the gases in the atmosphere.

UV radiation in the upper atmosphere causes the C–Cl bonds to break, producing chlorine radicals in the form of chlorine atoms. Chlorine atoms catalyse the decomposition of ozone, $2O_3 \rightarrow 3O_2$, increasing the rate of reaction by 15 000 times. This is a free radical chain reaction:

Initiation: $CF_2Cl_2 \rightarrow CF_2Cl\bullet + Cl\bullet$

Propagation: $Cl\bullet + O_3 \rightarrow ClO\bullet + O_2$

$ClO\bullet + O_3 \rightarrow 2O_2 + Cl\bullet$

The chlorine radical is regenerated and can destroy many more molecules of ozone. CFCs have a very long atmospheric lifetime (55 years) and this has caused so much ozone depletion that a hole in the ozone layer has been discovered.

Atmospheric data has been collected and analysed by different scientific research groups. This evidence has been used to create global legislation to ban the use of CFCs as solvents and refrigerants. Chlorine-free alternatives called HFCs (hydrofluorocarbons) were developed and these contain the stronger C–F bond. Research continues and the analysed data supports the conclusion that the ozone hole is repairing and should be fully restored by 2066.

✔ Summary

Ozone

1. Which of the following is a valid reason for most scientists believing that ozone in the upper atmosphere should not be allowed to become depleted? Tick **one** box. [1]

Ozone is an effective disinfectant ☐

Ozone absorbs wavelengths of UV radiation ☐

Ozone can naturally be decomposed but not naturally formed ☐

The ozone layer in the atmosphere helps to prevent global climate change ☐

2. Which statement is **not** true about ozone? Tick **one** box. [1]

Chlorine radicals catalyse the decomposition of ozone ☐

HFCs can catalyse the formation of ozone ☐

CFCs released into the atmosphere have led to the depletion of the ozone layer ☐

Ozone absorbs harmful radiation in the upper atmosphere ☐

3. The use of chlorotrifluoromethane ($CClF_3$) as a refrigerant is being phased out owing to concerns about ozone depletion in the upper atmosphere.

   a) Write an equation to show how $CClF_3$ can form a catalyst that increases the rate of decomposition of atmospheric ozone. Identify and name the catalyst. [2]

   _____

   _____

   b) Write equations to show the chain reaction that is catalysed by the species identified in part a). [2]

   _____

4. Bromotrifluoromethane ($CBrF_3$) is a halogenoalkane used in fire extinguishers in aircraft. Bromine atoms play a similar role to chlorine atoms in the decomposition of ozone.

   a) Write an equation for the formation of a bromine atom in the upper atmosphere. State the conditions needed for this reaction to happen. [2]

   _____

   _____

   b) A different halogenoalkane is chlorotrifluorimethane ($CClF_3$). $CClF_3$ is used as a refrigerant. Once used, halogenoalkane gases can be released into the air and reach the upper atmosphere.

   Suggest why it is more likely for $CBrF_3$ to produce bromine atoms than it is for $CClF_3$ to produce chlorine atoms. [1]

   _____

   _____

   c) Explain how bromine atoms ($Br\bullet$) act as a catalyst in the decomposition of ozone. Use equations in your answer. [5]

   _____

   _____

   _____

   _____

   _____

   _____

Alkenes

Spec. ref. 3.3.4.1

Key words

Key questions

What are alkenes?

What is the structure and bonding found in alkenes?

Why are alkenes likely to undergo reactions which include an electrophile?

What is the chemical test to show an organic substance is unsaturated?

Alkenes and their bonds

Alkenes are organic chemicals forming a homologous series that:

- are unsaturated hydrocarbons
- have a general formula of C_nH_{2n}.
- have the C=C functional group

There are strong covalent bonds between the atoms in alkenes. Each alkene has a simple covalent molecular structure with relatively weak, induced dipole-dipole forces between the molecules.

The C=C functional group

The functional group can lie in different positions in the main hydrocarbon chain. The different positions of the double bond result in **position isomers.**

But-1-ene   But-2-ene

The C=C functional group:

- is slightly shorter (120 pm) than the C–C functional group (154 pm)
- cannot rotate
- is planar with a bond angle of 120°.

If there are different groups bonded to each carbon atom in the functional group, **geometric isomers** can be formed.

E-but-2-ene   Z-but-2-ene

The C=C functional group has a high electron density, making alkenes susceptible to electrophilic attack with a carbocation reactive intermediate.

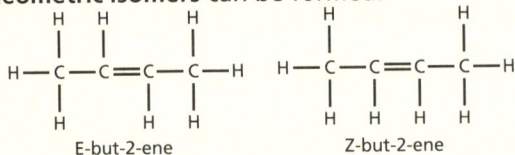

+ on the more substituted carbon

Testing for unsaturation

Organic compounds can be tested for unsaturation using bromine water, which is an orange-red colour. When bromine water is shaken with an unsaturated hydrocarbon, the bromine molecules 'add on' across the C=C double bonds and the reaction mixture quickly turns colourless, making a halogenated saturated organic compound.

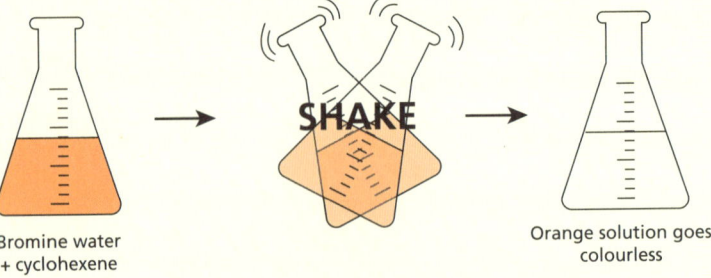

Bromine water + cyclohexene

SHAKE

Orange solution goes colourless

Summary

Spec. ref. 3.3.4.1

Alkenes

1 Which statement is **not** correct about $CH_2=C(CH_3)CH_3$? Tick **one** box. [1]

It decolourises bromine water ☐

It is likely to undergo electrophilic addition ☐

It is likely to undergo nucleophilic addition ☐

It cannot form geometric isomers ☐

2 Which statement about ethene is **correct**? Tick **one** box. [1]

Ethene forms a planar molecule ☐

Ethene contains polar covalent bonds ☐

Ethene has dipole-dipole forces between molecules ☐

Ethene forms position isomers ☐

3 The figure shows two stereoisomers of 4-methylpent-2-ene.

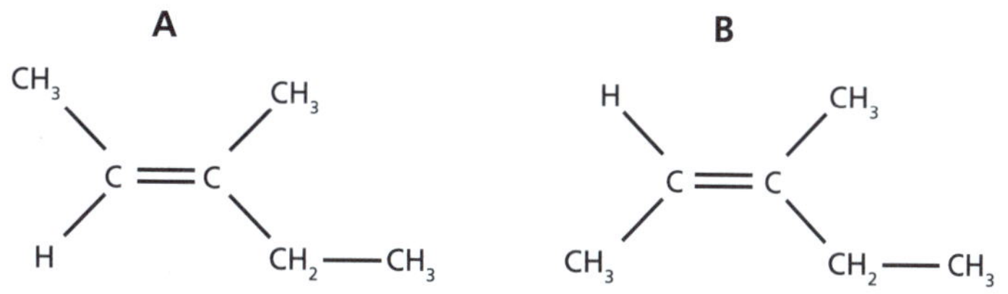

a) Determine which structure is Z-4-methylpent-2-ene. [1]

b) Explain why 4-methylpent-2-ene can show geometric isomers. [2]

c) Describe the chemical test to show that 4-methylpent-2-ene is unsaturated. Give the expected result. [3]

4 Hex-3-ene is a useful alkene intermediate in the synthesis of organic compounds.

a) State the meaning of the term 'alkene'. [2]

b) Draw the skeletal formula of E-hex-3-ene. [1]

c) Draw the structural formula of Z-hex-3-ene. [1]

Addition reactions of alkenes

🔑 Key words

❓ Key questions

What is an electrophile?

What is a carbocation?

Why do different carbocations have different stabilities?

Why is a mixture of products formed in the electrophilic addition reactions of some alkenes?

Electrophilic addition

Alkenes have a high electron density due to their functional group and they attract electrophiles. The C=C bond contains a pair of electrons that can be donated to form a covalent bond. Therefore, alkenes can undergo addition reactions where two reactants (an alkene and an electrophile) combine to form a single product.

Electrophilic addition of HBr

For symmetrical alkenes (e.g. ethene), only one product can be formed.

For asymmetrical alkenes (e.g. but-1-ene), there are two possible products in unequal amounts:

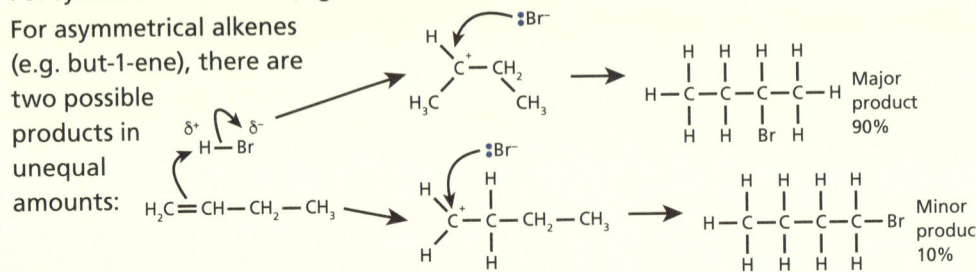

Stability of the intermediate carbocation determines which is the major product.

Stability of Carbocations

Electrophilic addition of H_2SO_4

When concentrated sulfuric acid is used, an alkyl hydrogensulfate functional group is formed. If water is added to the reaction mixture, then H_2SO_4 acts as a catalyst. This is a two-step mechanism where water is added across the alkene C=C bond to form an alcohol.

Step 1 (electrophilic addition):

Step 2 (hydrolysis):

ethyl hydrogensulfate   ethanol

Asymmetric alkenes produce an unequal mixture of products based on the stability of the carbocation reactive intermediate.

Electrophilic addition of Br_2

This reaction can be used as a test for saturation as bromine water decolourises as the halogen adds across the alkene C=C bond to form a colourless halogenoalkane.

✔ Summary

Addition reactions of alkenes

1. What is the IUPAC (International Union of Pure and Applied Chemistry) name of the major product formed from the electrophilic addition of hydrogen bromide to 2-ethylbut-1-ene? Tick **one** box. [1]

   2-bromo-2-methylpentane ☐

   2, 3-dibromomethylpentane ☐

   1-bromo-2-ethylbutane ☐

   3-bromo-3-methylpentane ☐

2. Which catalyst is used in the hydrolysis of an alkene? Tick **one** box. [1]

   Bromine water ☐

   Sulfuric acid ☐

   Water ☐

   Organic solvent, e.g. ethanol ☐

3. Two different products are formed when 2-methylbut-2-ene reacts with concentrated sulfuric acid.

   a) Give the name of the mechanism for this reaction. [1]

   b) Draw the **two** products of this reaction, identifying the major and minor product. [3]

   c) Explain why the products are formed in different amounts. [3]

4. 2-methylbut-2-ene can react with HBr to form a saturated organic compound.

   a) Give the name of the mechanism. [1]

   b) Draw the mechanism, identifying and naming the minor and major products. [6]

Addition polymers

🔲 (QR code)

🔑 Key words

❓ Key questions

What the relationship between monomers and polymers?

What is a repeating unit?

How are polymers named?

How can structure and bonding help you predict the properties of polymers?

Addition polymers and naming them

Addition polymers are made in a chemical reaction called polymerisation with 100% atom economy. Polymers are long-chain molecules made from smaller repeating units and they are unreactive.

The starting materials for a polymer are **monomers** and for addition polymers these must have a reactive C=C bond. So, addition polymers can be formed from alkenes and substituted alkenes.

Monomer

Polymer

Repeat unit

Polymers can have a brand name and a IUPAC name. The IUPAC naming convention for a polymer is the prefix 'poly' followed, in brackets, by the name of the monomer from which it is made. For example, Teflon is the brand name of a synthetic polymer that can be used in non-stick pans. The IUPAC name for Teflon is poly(tetrafluoroethene).

Properties

Changing the conditions of polymerisation or changing the monomer affects the properties of the polymer. This knowledge has developed over time as:

- scientists experimented and collected data
- technology advanced, bringing new catalysts and different processes to control the conditions more accurately
- new raw materials were discovered.

Addition polymers are unreactive due to the bonding in the carbon chain, which is:

- saturated and so a lot of energy is needed to break the strong C–C covalent bonds
- non-polar and therefore not susceptible to nucleophilic or electrophilic attack.

The unreactive nature of addition polymers makes them durable and non-biodegradable.

The stronger the intermolecular force, the higher the melting point of the polymer. Here are some examples:

The melting point of poly(ethene), which only contains induced dipole-dipole forces, is about 120°C.	As poly(chloroethene) contains a polar C–Cl bond, it has both induced dipole-dipole forces and permanent dipole-dipole forces. More energy is needed to overcome the forces and so it has a higher melting point of 165°C.	Poly(ethenol) contains both induced dipole-dipole forces and hydrogen bonds between the polymer chains and as hydrogen bonds are about 10% the strength of a covalent bond, this leads to an even higher melting point of 230°C.

✔ Summary

Addition polymers

1 Which statement about poly(propene) is correct? Tick **one** box. [1]

It decolourises bromine water ☐

It has a lower relative molecular mass than propene ☐

It has a higher melting point than propene ☐

It has permanent dipoles between the polymer chains ☐

2 Which polymer chain has hydrogen bonding between its chains? Tick **one** box. [1]

Poly(buten-1-ol) ☐   Poly(dichloropropene) ☐

Poly(2-methylbut-2-ene) ☐   Poly(tetraiodoethene) ☐

3 Nonane can be cracked to form large quantities of propene.

   a) Name the reaction that can produce poly(propene) from propene. [1]

   b) Draw the repeating unit of poly(propene). [1]

   c) Describe the structure and bonding in poly(propene). [4]

4 Halon is one brand name for the addition polymer shown below.

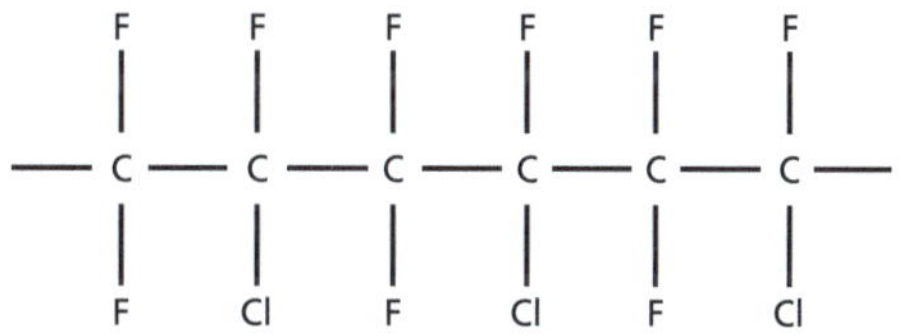

   a) Give the IUPAC name for this polymer. [1]

   b) Draw the monomer which would be used to make this polymer. [1]

   c) Suggest how the melting point of this polymer would compare to the melting point of poly(ethene). [4]

Alcohols

Key words

Key questions

What are the two ways that ethanol can be industrially produced?

What is a biofuel?

What does 'carbon neutral' mean?

What are the environmental (including ethical) issues linked to decision making about biofuel use?

Alcohols and the production of ethanol

Alcohols are a homologous series with the –OH functional group. Ethanol, CH_3CH_2OH, is an example of an alcohol that has medicinal and industrial uses.

Direct hydration of ethene requires 300°C, 70 atmospheres pressure and a concentrated phosphoric acid catalyst. The ethene is obtained from the fractional distillation of crude oil and then steam cracking of the long-chain hydrocarbons.

$$CH_2 = CH_2 + H_2O \xrightarrow[\text{catalyst}]{\text{phosphoric acid}} C_2H_5OH$$

Fermentation uses the enzyme, zymase, which is produced by yeast. The optimum conditions for the reaction are anaerobic and a temperature of 30–40°C.

$$C_6H_{12}O_6 \text{ (aq)} \rightarrow 2C_2H_5OH \text{ (aq)} + 2CO_2 \text{ (g)}$$

The product must be purified by fractional distillation as it is an aqueous solution of about 15% ethanol.

Manufacture method / Measure	Direct hydration	Fermentation
Sustainability	Finite resources used	Renewable resources used (e.g. waste plant material)
Atom economy	100%	51%
Rate of reaction	Fast	Slow
Type of process	Continuous	Batch
Energy	High as high temperature and pressure	Lower energy requirement as needs moderate temperatures

Biofuels

Biofuels are renewable energy resources made from living organisms that can be used as a hydrocarbon fuel replacement. Biofuels are often considered to be **carbon neutral**. This means that the amount of carbon taken in by the organism is equal to the amount of carbon released when the biofuel is used. However, this idea often doesn't take into account the carbon released during the growing of the organic material.

Biofuels can help to reduce the carbon footprint for the combustion of fuels. However, this can:

- lead to deforestation in order to increase the capacity to grow energy crops
- lead to changes in land use, contributing to reduced yields of food crops, price rises and food supply issues.

✔ Summary

Alcohols

1 Which statement is **not** correct about the industrial production of ethanol from ethene at 300°C? Tick **one** box. [1]

The atom economy is 100% ☐ The reaction is catalysed by zymase ☐

A higher pressure than atmospheric pressure is used ☐ Pure ethanol is produced ☐

2 Which statement is correct about the production and use of ethanol as a biofuel? Tick **one** box. [1]

Bioethanol is purified by cracking ☐

During combustion reactions, bioethanol does not produce oxidised carbon ☐

Bioethanol burns more completely that ethanol ☐

Bioethanol is made from anaerobic fermentation with yeast ☐

3 Petrol in the UK contains 10% bioethanol. Bioethanol is ethanol made from crops and is considered to be carbon neutral.

a) Name the reaction used to make bioethanol from crops. [1]

b) State what is meant by the term 'carbon neutral'. [1]

c) Justify bioethanol as being carbon neutral. Use balanced equations in your answer. [4]

4 Ethanol can be made in industry by a continuous process. The reaction is an addition reaction between ethene and steam.

a) Explain from where ethene is obtained for this industrial process. [3]

b) Use curly arrows to complete the reaction mechanism for the direct hydration of ethene. [3]

c) Name the catalyst used. [1]

Reactions of alcohols

🔲 (QR code)

🔑 Key words

❓ Key questions

What does [O] mean?

What are the products of the oxidation of a primary alcohol?

What is the product of the oxidation of a secondary alcohol?

What chemical tests can be used to distinguish between aldehydes and ketones?

Write down a use for the organic product from the elimination of water from ethanol.

Classifying alcohols

Alcohols are organic molecules that contain the −OH functional group. The position of the functional group can be used to classify an alcohol.

Primary alcohol   Secondary alcohol   Tertiary alcohol

Oxidation of alcohols

The removal of the hydrogen atom from the −OH functional group occurs during the oxidation of alcohols. Acidified potassium dichromate(VI) is a suitable oxidising agent that is often simplified to [O], meaning oxidising agent. The reaction mixture is heated under reflux and the orange of the dichromate becomes green as the CrO_4^{2-} is reduced to form Cr^{3+}.

Primary alcohols oxidise first to aldehydes and then to carboxylic acids:

$CH_3CH_2OH \xrightarrow{[O]} CH_3CHO \xrightarrow{[O]} CH_3COOH$

ethanol   oxidation (alcohol in excess – no reflux)   ethanal (an aldehyde)   oxidation (oxidising agent in excess – reflux)   ethanoic acid (a carboxylic acid)

Secondary alcohols oxidise to form ketones when heated under reflux:

$CH_3CH(OH)CH_3 \xrightarrow{[O]} CH_3COCH_3 + H_2O$

propan-2-ol   propanone (a ketone)

Tertiary alcohols do not easily undergo oxidation.

Chemical tests

Aldehydes can be oxidised further to make carboxylic acids, but ketones cannot. Some transition metal ions can be reduced by aldehydes, but not by ketones, and this can be used to generate a practical test to distinguish between the carbonyl-containing compounds.

Tollens' reagent is colourless ammoniacal silver nitrate with the reactive species $[Ag(NH_3)_2]^+$. When warmed with aldehydes, a silver precipitate is formed and a mirror is seen. There is no observable change with ketones.

Fehling's solution contains blue Cu^{2+} ions which, when heated with an aldehyde, is reduced to form a brick red precipitate of copper metal. There is no observable change with ketones.

Elimination reaction

Alcohols can be dehydrated to produce an alkene and a water molecule when heated with concentrated sulfuric acid or concentrated phosphoric(V) acid, in the presence of an aluminium oxide catalyst.

$+ H_2O + H^+$

The alkenes produced this way can be used to make addition polymers without using crude oil.

✔ Summary

Reactions of alcohols

1. Which of the following is a catalyst for the dehydration of ethanol? Tick **one** box. [1]

   Potassium dichromate ☐

   Aluminium oxide ☐

   Concentrated sulfuric acid ☐

   Dilute sulfuric acid ☐

2. What happens when ethanal is mixed with warmed Fehling's solution? Tick **one** box. [1]

   Colourless solution turns brick red ☐

   Blue solution turns brick red ☐

   Colourless solution forms a silver mirror ☐

   Blue solution forms a silver mirror ☐

3. Ethan-1,2-diol is an organic substance that contains two –OH on each carbon atom. This diol can undergo an oxidation reaction with concentrated potassium dichromate to make an organic acid.

   a) Suggest the conditions required for this reaction to occur. [2]

   b) Name the product of this reaction. [1]

   c) Justify your classification of ethan-1,2-diol as a primary, secondary or tertiary alcohol. [2]

4. Chromatography was used to separate a mixture of pentan-2-ol and 2-methylbutan-2-ol.

   a) Describe a chemical test to identify each alcohol. Give the reagents and describe what would be observed for each alcohol. [3]

   b) Pentan-2-ol can undergo a dehydration reaction to produce a mixture of organic products.

   i) Name the **two** organic products. [2]

   ii) Describe the reaction conditions for the dehydration of pentan-2-ol. [3]

Identification of functional groups

Key words

Key questions

What information should you note from observing simple test tube reactions?

What is a functional group?

How can you identify which halogen is present in a halogenoalkane?

Why can acidified potassium dichromate not be used to distinguish between an aldehyde and a primary alcohol?

Observing test tube reactions

When studying organic substances at room temperature, bear in mind that:

- organic solids suggest high molecular mass.
- organic liquids suggest medium molecular mass with induced dipole-dipole forces or smaller carbon chains with polar bonds that lead to permanent dipole-dipole forces or hydrogen bonds.
- organic gases suggest low molecular mass, i.e. non-polar with induced dipole-dipole forces between the molecules.

When organic solvents like hexane are used, the organic substance is likely to be non-polar. If ethanol or water is used, the organic substance is likely to have a dipole.

Practical tests

Functional group	Description of the test	Observation
$C=C$ (alkenes)	Shake with bromine water	Bromine water decolourises if the organic substance is unsaturated.
$-COOH$ (carboxylic acids)	1. Use universal indicator paper or a pH probe. 2. React with sodium hydrogen carbonate solution ($NaHCO_3$).	1. pH less than 7. 2. Effervesces and when the gas is blown through limewater, the limewater changes from colourless to cloudy.
$C=O$ (to distinguish between aldehydes and ketones)	1. Warm with Fehling's solution. 2. Warm with Tollens' reagent. 3. Warm with acidified potassium dichromate ($K_2Cr_2O_7$).	1. With an aldehyde, a blue solution forms a brick red precipitate. No observable change with a ketone. 2. With an aldehyde, a silver precipitate forms, leaving a silver mirror on the inside of the test tube. No observable change with a ketone. 3. With an aldehyde, the orange solution turns green. No observable change with a ketone.
$C-X$ (halogenoalkanes)	Warm a mixture of the halogenoalkane with dilute sodium hydroxide (NaOH). Acidify the solution with some dilute nitric acid (HNO_3), then add some silver nitrate solution ($AgNO_3$).	The colour of a precipitate indicates the halogen present – white for chlorine, cream for bromine and yellow for iodine. To distinguish the colour further, the reaction of the precipitate with ammonia can be studied: • AgCl dissolves in dilute and concentrated ammonia solution. • AgBr dissolves only in concentrated ammonia solution. • AgI is insoluble in ammonia solution.
$-OH$ (alcohol)	Add acidified potassium dichromate ($K_2Cr_2O_7$)	With primary and secondary alcohols, the orange solution turns green. No observable change with tertiary.

Summary

Identification of functional groups

1 Which reagent can be used to distinguish between pentanal and butan-2-one? Tick **one** box. [1]

Acidified potassium dichromate ☐

Bromine water ☐

Acidified silver nitrate ☐

Sodium hydrogen carbonate solution ☐

2 Which hydrocarbon decolourises bromine water in the absence of UV light? Tick **one** box. [1]

$CH_2CHCH_2CH_2CH_2CH_3$ ☐

$CH_3CH_2CH_2CH_2CH_2CH_3$ ☐

$CH_3CH(CH_3)(CH_2)_2CH_3$ ☐

$CH_2C(OH)CH_2CH_2CH_2CH_3$ ☐

3 Two different colourless liquids are known to be organic chemicals with the molecular formula $C_4H_8O_2$.

a) Both organic substances effervesce with a solution of sodium hydrogen carbonate.

Give the functional group present. [1]

b) Suggest the pH of the organic liquids. [1]

c) Name the **two** possible organic substances. [2]

4 The structures of two organic substances are shown.

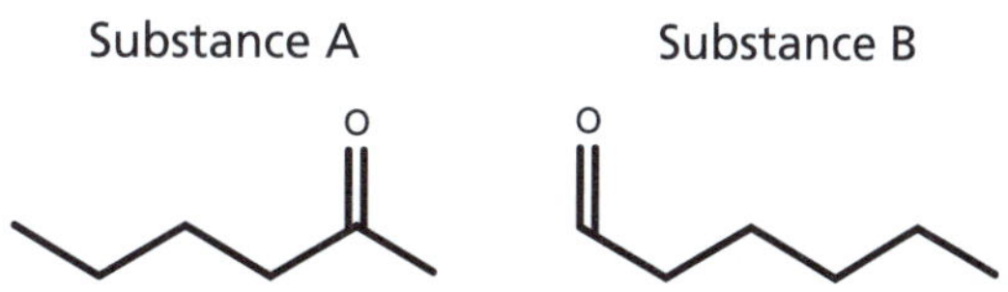

Substance A   Substance B

a) Explain the relationship between substance A and substance B. [2]

b) Describe a simple test tube reaction that could be used to distinguish between substance A and substance B. Give the expected observations in your answer. [4]

Organic chemistry **109**

Mass spectrometry

What information can mass spectrometry give?

What is a molecular ion?

What is a base peak?

How can mass spectrometry give an indication of the structure of an organic substance?

Mass spectrometry

Mass spectrometry is an analytical technique that can be used to determine the:
- isotopic ratio of an element
- molecular mass of a substance
- identity of an unknown substance.

A small sample of the substance is put into a mass spectrometer. The process may cause organic molecules to be broken down into smaller parts by random bond fission.

The mass spectrometer detects the molecular ion and any of these fragment ions. This data, along with other spectroscopic techniques and simple test tube analysis, can be used to determine the structure of organic substances.

m/z	Possible ion
15	$[CH_3]^+$
29	$[CH_3CH_2]^+$
43	$[CH_3CH_2CH_2]^+$ $[CH_3CO]^+$
77	$[C_6H_5]^+$

Identifying peaks

The sum of the abundance of the peaks for an element is 100% but for a molecule this is not the case. The molecular ion is formed by: $M \rightarrow M^{+\bullet} + e^-$

This appears on the spectrum as the largest peak with the highest m/z ratio. There is usually a M+1 peak due to the presence of the ^{13}C isotope. This peak gets larger as the number of carbon atoms in the organic compound increases (as the probability of the molecule containing one ^{13}C isotope increases).

The base peak is the ion with the highest relative abundance. It may not be the molecular ion. This fragment is the most stable ion in the mass spectrometer and can be useful for identifying the compound.

For example, molecule X is an organic substance containing a C=O functional group. Using the mass spectrum shown below, you can see that the $M_r = 72$. The base peak occurs at m/z 43, which is likely to be CH_3CO^+. There is a major peak at m/z = 29, which is likely to be $CH_3CH_2^+$. These data can be used to determine that x is butanone.

Mass spectrum of molecule butanone

Relative intensity

Base peak

Molecular ion

10 15 20 25 30 35 40 45 50 55 60 65 70 75
m/z

Additional information, such as test tube reactions of a negative result with both Fehling's solution and Tollens' reagent, suggest that this substance is a ketone.

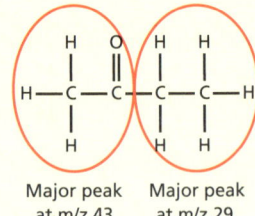

Major peak at m/z 43   Major peak at m/z 29

Summary

Mass spectrometry

(1) A high-resolution mass spectrum showed a molecular ion peak at 44.0261

Which is the correct molecular formula of this compound? Tick **one** box. [1]

CO_2 ☐ C_2H_4O ☐ C_3H_8 ☐ C_4H_8 ☐

(2) The mass spectrum of an aldehyde has a major fragment at m/z = 31.

What is the formula of this fragment? Tick **one** box. [1]

CH_3O^+ ☐ CH_3O ☐ $CH_3O^•$ ☐ $^+CH_2OH$ ☐

(3) A sample of propanol was analysed using a mass spectrum.

The table shows some precise relative atomic mass values.

Atom	Precise relative atomic mass
^1H	1.00794
^{12}C	12.0000
^{16}O	15.9949

a) State why the precise relative atomic mass for the ^{12}C isotope is exactly 12.00000 [1]

b) Use the data to calculate the precise molecular mass of propanol. Give your answer to the same precision as the data in the table. [2]

c) Suggest a reason for a small peak on the mass spectrum at m/z = 61. [2]

(4) An organic compound has the molecular formula C_4H_7ClO. It is a volatile liquid that does not produce misty fumes when added to water. The compound is analysed using mass spectrometry.

a) State the meaning of the term 'molecular ion'. [1]

b) Suggest why the mass spectrum of this organic compound contains two molecular ion peaks at m/z = 106 and m/z = 108. [1]

c) Identify the fragment ion at m/z = 43, which contains atoms of three different elements. [1]

Infrared spectrometry

Key words

Key questions

Where is the fingerprint region in an IR spectrum?

Why is the fingerprint region useful in an IR spectrum?

How can an IR spectrum help to identify the structure of a molecule?

What is the effect of greenhouse gases in Earth's atmosphere?

Infrared (IR) spectroscopy

A sample is loaded into a machine and IR light is passed through the sample. Some bonds will absorb IR radiation above 1500 cm^{-1} and these characteristic absorptions can be used to identify bonds in a sample.

Decoding IR spectra

The fingerprint region is below 1500 cm^{-1} and this is the part of the spectrum that is used to identify the molecule by comparison against other known spectra.

You will be given a table in the data sheet of your A-level exam paper to help identify the presence of certain functional groups (see Table A, page 219).

For example, the IR spectrum of propanoic acid has a characteristic fingerprint region below 1500 cm^{-1}. The bonds that are part of the functional group (–COOH) have characteristic absorptions and can be identified.

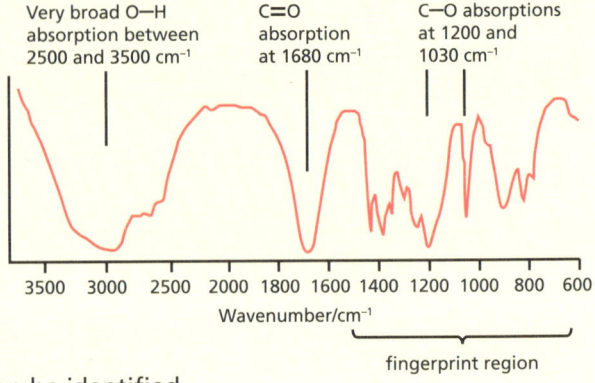

Unexpected absorptions can occur due to impurities in the sample, such as solvents used in the sample preparation.

Global warming

Sunlight enters the Earth's atmosphere as short-wave UV radiation. Some of this radiation is reflected from the Earth's surface as shorter wave IR radiation. The bonds in the greenhouse gases (CO_2, H_2O and CH_4) absorb the reflected IR radiation from the surface of the Earth. This is called the **greenhouse effect** and is important in keeping global temperatures high enough to support life on Earth.

However, as the concentration of greenhouse gases increases in the atmosphere, more IR radiation is absorbed and this causes a heating effect called **global warming**. Global warming is thought to be the main contributor to global climate change.

The total absorption and scattering considers all the gases in the atmosphere, not only the main greenhouse gases. Notice how the large absorption for carbon dioxide occurs in the IR region.

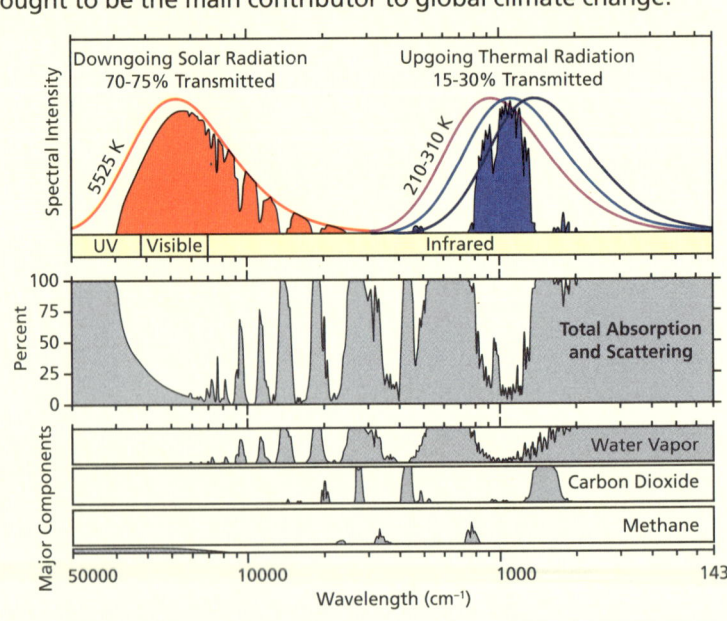

Summary

Infrared spectrometry

1 Which of the following atmospheric gases do **not** absorb infrared radiation? Tick **one** box. [1]

CO_2 ☐ O_2 ☐ H_2O ☐ CH_4 ☐

2 The infrared spectrum of an organic compound is shown.

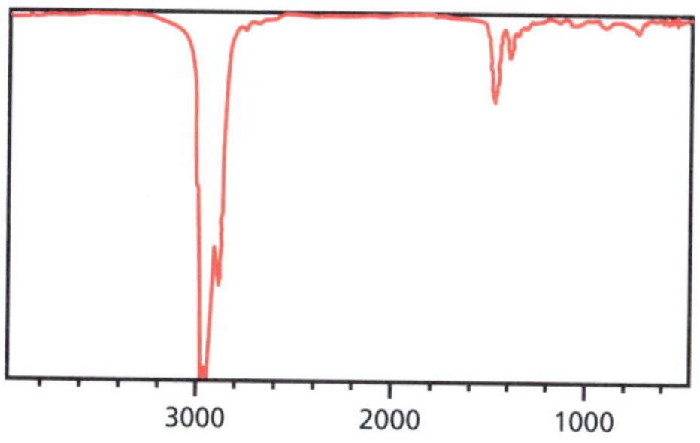

Wavenumber/cm^{-1}

Which compound could produce this spectrum? Tick **one** box. [1]

Hexane ☐   Hexan-1-ol ☐

Hexanoic acid ☐   Ethylhexanoate ☐

3 This question is about an organic compound with the molecular formula $C_4H_8O_2$. The infrared spectrum for this compound is shown.

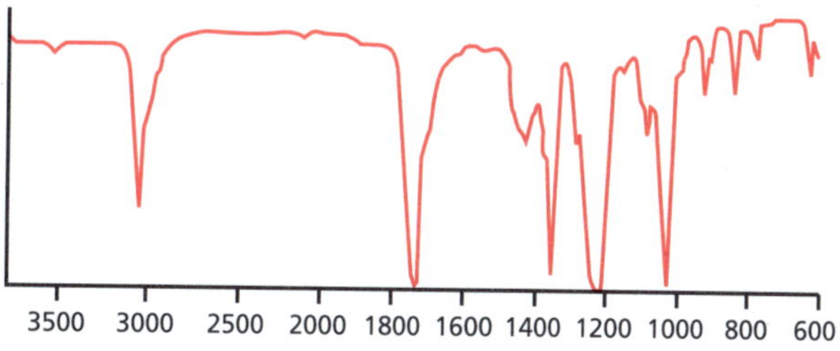

Wavenumber/cm^{-1}

a) Identify the bond that causes the absorption at 1740 cm^{-1}. [1]

b) Suggest the functional group present in this organic molecule. [1]

c) Explain how the infrared spectrum could be used to identify the organic compound. [2]

4 Butanal has the molecular formula C_4H_8O. There are seven other isomers that have the same molecular formula.

a) Name the functional group isomer of butanal that has an absorption in the range 1680–1750 cm^{-1} in its infrared spectrum. [1]

b) i) Draw the skeletal formula of a saturated structural isomer of butanal that has an absorption in the range 3230–3550 cm^{-1} in its infrared spectrum. [2]

ii) Name the isomer. [1]

Optical isomerism

Key words

Key questions

What is optical isomerism?

Isomerism

Isomers are different substances that have the same molecular formula. Stereoisomerism is where the molecular and structural formulae are the same but there is a different arrangement of atoms in space.

Isomerism
- Structural
  - Chain
  - Positional
  - Functional
- Stereo
  - Geometric, e.g. E/Z — **Isomers differ in their spatial arrangement about a double bond.**
  - Optical — **Isomers differ in the arrangement of atoms in 3D space which create mirror images of each other.**

Optical isomerism

Optical isomers are pairs of molecules that are non-superimposable mirror images. They have similar chemical and physical properties, but different biological properties. Complex ions can form optical isomers as they form two different structures that are mirror images and non-superimposable.

Chiral centres are often carbon atoms and are shown in molecules using an asterisk (*).

What is a chiral centre?

2-hydroxypropenoic acid

A chiral centre is an asymmetric carbon atom with four different groups on it and this allows them to be arranged differently and form non-superimposable mirror images known as **enantiomers**.

Isomers are given the prefix + if they rotate the light clockwise and the prefix – if they rotate the light anticlockwise. Limonene is an organic substance found naturally in plants. The chiral centre allows limonene to form enantiomers. They can be distinguished by using a plane of polarised light but they smell differently too.

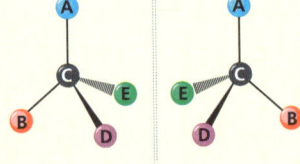

Optical Isomerism

Imaginary Mirror

What effect do enantiomers have on polarised light?

(+) - Limonene
Rotates polarised light clockwise
Smells like oranges

(–) - Limonene
Rotates polarised light anticlockwise
Smells like pine

A mixture of equal amounts of enantiomers is called a **racemic mixture** (racemate). As the enantiomers are in equal proportion, there is no overall effect on polarised light.

✔ Summary

Optical isomerism

(1) Which of the following statements is **not** true about 1,2-dichloropropene? Tick **one** box. [1]

1,2-dichloropropene has an unsymmetrical carbon atom ☐

1,2-dichloropropene has a chiral centre ☐

A racemic mixture of 1,2-dichloropropene rotates polarised light ☐

1,2-dichloropropene shows E/Z isomerism ☐

(2) Which one of the following cannot produce optically active isomers? Tick **one** box. [1]

Reduction of 2,3-dimethylpent-2-ene with H_2 in the presence of a nickel catalyst ☐

Addition reaction between HCl and but-2-ene ☐

Reduction of $CH_3CH_2COCH_3$ ☐

Dehydration of butan-2-ol by heating with concentrated sulfuric acid ☐

(3) An organic compound shows optical activity and has the molecular formula $C_3H_6Cl_2$

  a) Draw the structural formula of the organic compound and identify the chiral carbon. [2]

  b) The organic compound can form enantiomers.

  Define the term 'enantiomer'. [2]

  c) Explain why a racemic mixture has no effect on polarised light. [3]

(4) Alanine is an amino acid and can exist as a pair of stereoisomers.

$$H_2N - \underset{\underset{H}{|}}{\overset{\overset{CH_3}{|}}{C}} - COOH$$

  a) Define the term 'stereoisomer'. [2]

  b) Describe how to distinguish between the two stereoisomers. [2]

Carbonyls

Key words

Key questions

What are carbonyl-containing compounds?

Why are carbonyl compounds susceptible to nucleophilic attack?

What is formed if an aldehyde is reduced?

What is formed if a ketone is reduced?

How are hydroxynitriles produced?

The carbonyl functional group

The carbonyl functional group (C=O) can be found in aldehydes, ketones, and in carboxylic acids and their derivatives. The electronegativity of the oxygen atom causes a dipole across the carbonyl group and this makes it susceptible to nucleophilic attack.

Aldehydes

Aldehydes can easily be oxidised to carboxylic acids, but ketones cannot. An aldehyde can be summarised as RCHO and an oxidant can be summarised as [O]. Therefore, the general equation for the oxidation of an aldehyde is: $RCHO + [O] \rightarrow RCO_2H$

This fact is used in simple test tube reactions (using Fehling's solution or Tollens' reagent) to distinguish between aldehydes and ketones.

Aldehydes can be reduced to primary alcohols using $NaBH_4$ in aqueous solution. The $NaBH_4$ is the reductant and a source of the hydride ion, H^-. Reductants are often summarised as [H] in equations.

The general equation for the reduction of an aldehyde to make a primary alcohol is: $RCHO + 2[H] \rightarrow RCH_2OH$ (This reaction can be classified as both a reduction and a nucleophilic addition, with the nucleophile being a hydride ion, H^-.)

Ketones

Ketones are not oxidised by Fehling's solution or Tollens' reagent. However, ketones can be reduced to secondary alcohols using $NaBH_4$ in aqueous solution. This reaction can be classified as both a reduction and a nucleophilic addition.

Hydroxynitriles

Hydroxynitriles are organic compounds that contain a cyano (–CN) and a hydroxy (–OH) group attached to the same carbon atom.

This organic compound, 2-hydroxy-2-methylpropanenitrile, is an example of a hydroxynitrile.

Under room temperature and pressure, carbonyl compounds can undergo nucleophilic addition reactions with KCN, followed by dilute acid. In the first step of the mechanism, the KCN supplies CN^-; this is a toxic ion and KCN will fully ionise, giving a greater concentration than if HCN were used. In the second step, the H_2SO_4 gives H^+. This reaction can be used to extend the carbon chain length.

Aldehydes and unsymmetrical ketones form mixtures of enantiomers when they react with KCN followed by dilute acid.

Summary

Carbonyls

1. Which reagents are needed to form a racemic mixture with CH_3CHO? Tick **one** box. [1]

KCN and dilute H_2SO_4 ☐

$K_2Cr_2O_7$ with concentrated H_2SO_4 ☐

Acidified $KMnO_4$ ☐

A solution made from $AgNO_3$, NH_4OH and NaOH ☐

2. Which alcohol could **not** be produced by the reduction of an aldehyde or a ketone? Tick **one** box. [1]

2,3-dimethylbutan-1-ol ☐

2,3-dimethylpent-3-ol ☐

2-methylbutan-2-ol ☐

2-methylpropan-2-ol ☐

3. The diagram shows the first step of a mechanism to produce 2-hydroxy-2-methylpropanenitrile.

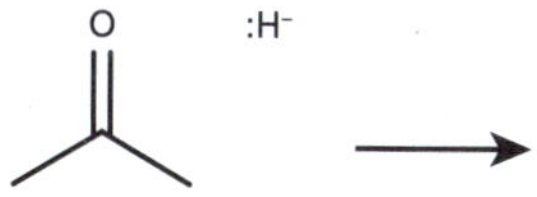

a) Complete the reaction mechanism above. [3]

b) Give the name of the reaction mechanism. [1]

c) Name and describe the functional group present in the product of this reaction. [3]

4. The diagram shows three organic compounds, S, T and U, with the molecular formula $C_6H_{12}O$.

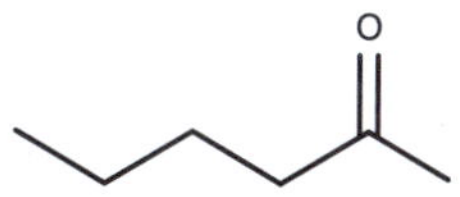

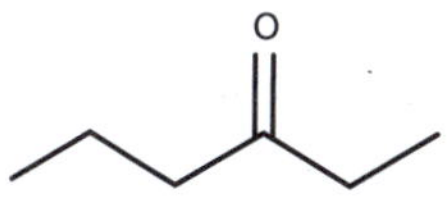

a) Name the type of isomerism shown between S, T and U. [1]

b) i) Identify the isomer(s) that would react with Tollens' reagent. [1]

   ii) State the expected observation when Tollens' reagent reacts. [1]

c) i) Identify the isomer(s) that would react when warmed with acidified potassium dichromate(VI). [1]

   ii) State the expected observation when acidified potassium dichromate(VI) reacts. [1]

d) Identify the isomer(s) that would give a mixture of enantiomers when reacted with KCN, followed by dilute acid. [1]

Carboxylic acids and esters

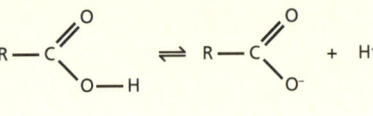

🔑 Key words

❓ Key questions

Why are carboxylic acids weak acids?

How are esters made?

What is biodiesel?

What is soap?

Carboxylic acids

Carboxylic acids (also called organic acids) have the functional group –COOH. They are weak acids as they partially ionise in aqueous solution to release protons, H^+ (aq).

$$R-C \underset{O-H}{\overset{O}{\big\langle}} \rightleftharpoons R-C \underset{O^-}{\overset{O}{\big\langle}} + H^+$$

Carboxylic acids react with:
- reactive metals to form a metal salt (metal carboxylate) and hydrogen gas.
- carbonates to form a metal salt, water and carbon dioxide (this can be used as a simple test tube reaction to identify a carboxylic acid).
- alcohols in the presence of concentrated strong acid catalyst will undergo a reversible reaction to form esters and water, e.g.

$$CH_3COOH + CH_3CH_2OH \xrightarrow{H^+} CH_3COOC_2H_5 + H_2O$$

This reaction takes place with heating under reflux.

Carboxylic acids cannot be oxidised. The exception is methanoic acid, which can be oxidised to form carbonic acid. Carbonic acid can undergo a reversible reaction to decompose into carbon dioxide gas and water.

Esters

Esters have the functional group –COO–

They have many common uses including in solvents, plasticisers, perfumes and food flavourings.

$$R''-C \overset{O}{\overset{\|}{}} -O-R'$$

Esters can be hydrolysed in acidic or alkaline conditions to form alcohols and carboxylic acids or salts of carboxylic acids. The reaction takes place when heated under reflux. Under acidic conditions yields are low, but under excess alkaline conditions the reaction goes to completion.

$$CH_3COOCH_3 + H_2O \xrightarrow{H^+} CH_3COOH + CH_3OH$$

$$CH_3COOCH_3 + H_2O \xrightarrow{OH^-} CH_3COO^- + CH_3OH$$

Vegetable oils and animal fats are esters of propane-1,2,3-triol (glycerol) and long-chain fatty acids. Vegetable oils and animal fats can be hydrolysed in alkaline conditions to give soap, which is made of salts of long-chain carboxylic acids and glycerol.

Biodiesel is an alternative to the mineral diesel made from the fractional distillation of crude oil. It is a mixture of methyl esters of long-chain carboxylic acids and made by reacting vegetable oils with methanol in the presence of a strong alkali catalyst.

Biodiesel is often considered to be carbon neutral and vegetable oil can be re-used from waste sources, such as fish-and-chip shops, to help reduce waste.

✔ Summary

Carboxylic acids and esters

1 Which reagents are needed to form an ester with ethanoic acid? Tick **one** box. [1]

Ethanol and dilute sulfuric acid ☐

Ethanol and concentrated sulfuric acid ☐

Ethene and nickel ☐

Sodium hydrogen carbonate and ethanol ☐

2 Which one of the following is **not** correct about vitamin C, shown below? Tick **one** box. [1]

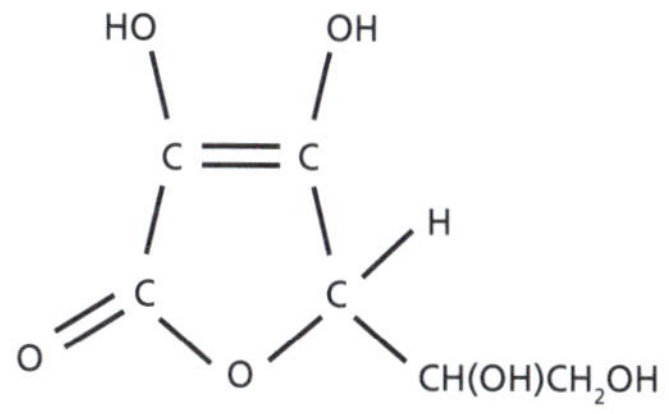

Vitamin C is a cyclic ester ☐

Vitamin C is unsaturated ☐

Vitamin C is an oxidant ☐

Vitamin C is a reductant ☐

3 The triglyercide of hexadecanoic acid can undergo hydrolysis in the presence of NaOH.

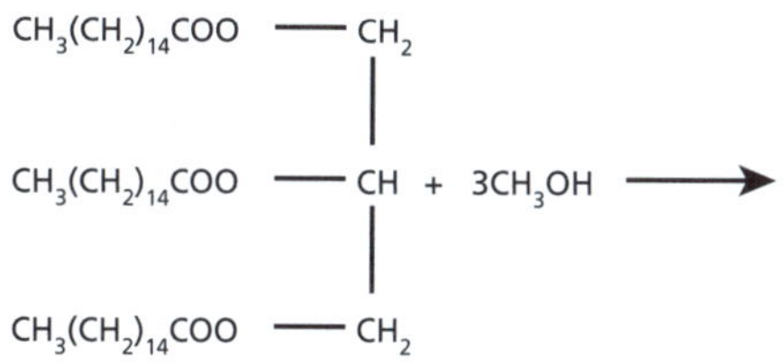

a) Complete the equation for this reaction. [2]

b) Describe the role of the NaOH. [1]

c) Give **one** use for the product of this reaction. [1]

4 Pineapple flavouring in sweets can be achieved by using a food additive made from ethyl butanoate.

a) Write an equation for the preparation of ethyl butanoate from an acid and an alcohol. [3]

b) Give a catalyst and the reaction conditions used for the laboratory scale preparation of ethyl butanoate. [3]

c) i) Identify the functional group present. [1]

ii) Name the homologous series that the organic product belongs to. [1]

Acylation

Key words

Key questions

Why is a carboxylic acid derivative?

What is acylation?

What sort of reactions can acyl chlorides and acid anhydrides undergo?

What are the industrial advantages of using ethanoic anhydride to manufacture aspirin?

Derivatives of carboxylic acids and uses of acylation reactions

When an atom, or a group of atoms, is replaced in a carboxylic acid functional group (–COOH) then carboxylic acid derivatives are made.

In an **acylation reaction**, an acyl group (R–C=O) is added to a compound.

Aspirin (2-acetoxybenzoic acid) is a drug used to treat pain. It is made by the reaction between a carboxylic acid and an acid anhydride. Ethanoic anhydride is used in preference to ethanoyl chloride as it is safer (less corrosive and the by-products are less dangerous), it is cheaper, it doesn't readily react with water and reactions are more easily controlled.

Reactions of acyl chlorides and acid anhydrides

Water – a carboxylic acid is produced	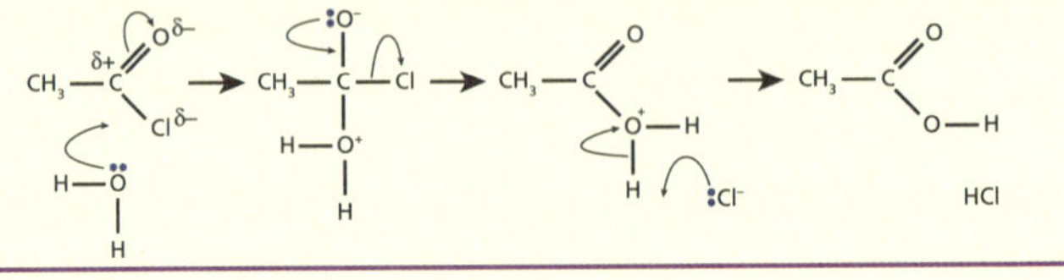
Alcohol – an ester is produced	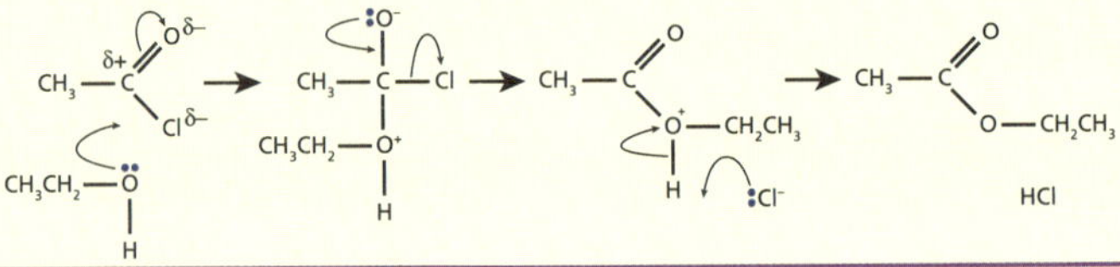
Primary amines – an N-substituted amide is produced by either of the following mechanisms	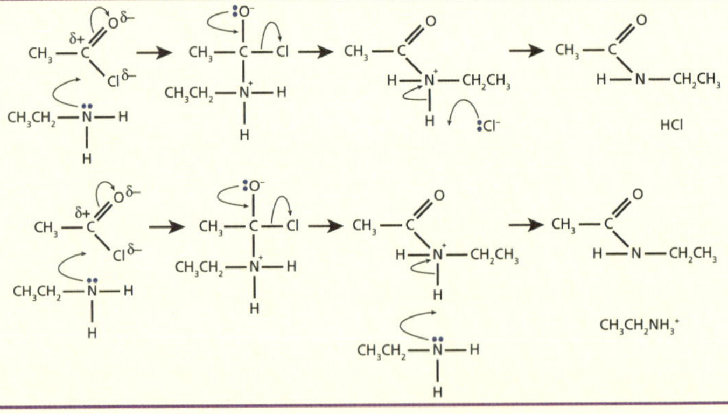
Ammonia – an amide is produced	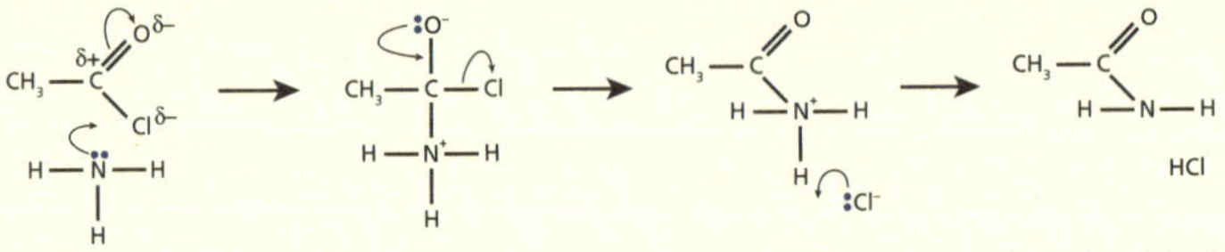 N.B. In the last step of the mechanism, the hydrogen can be removed by the chloride ion or an ammonia molecule.
Water – a carboxylic acid is produced	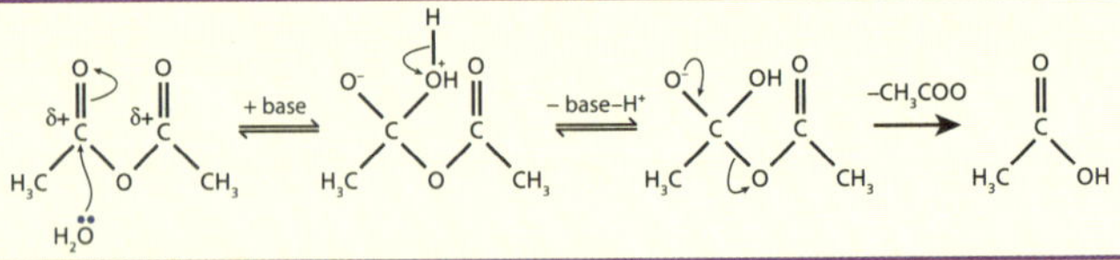
Alcohol – an ester is produced	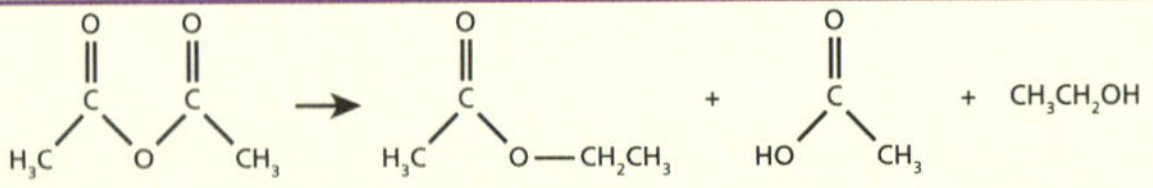
Primary amines – an N-substituted amide is produced	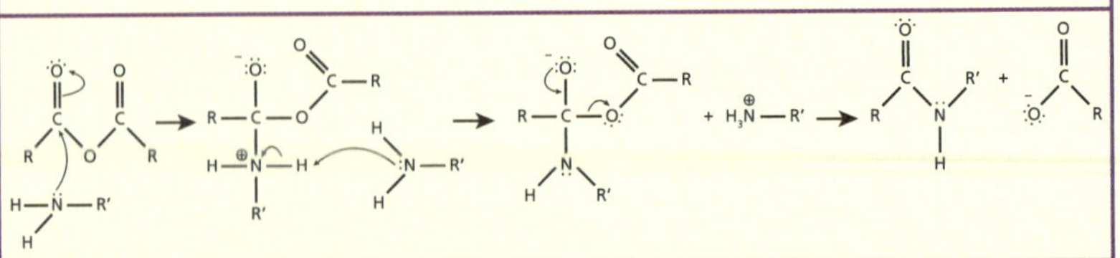

Summary

Acylation

1 Which of the following is **not** a carboxylic acid derivative? Tick **one** box. [1]

Acyl chloride ☐

Acid anhydride ☐

N-substituted amide ☐

Primary alcohol ☐

2 What reaction mechanism happens when ethanoyl chloride reacts with ethanol? Tick **one** box. [1]

Nucleophilic addition-elimination ☐

Electrophilic addition-elimination ☐

Free radical ☐

Nucleophilic substitution ☐

3 Aspirin can be prepared by acylation using either ethanoyl chloride or ethanoic anhydride.

a) Complete the equations shown below.

i) $CH_3COCl + HOC_6H_4COOH \rightarrow$ .... [1]

ii) $(CH_3CO)_2O + HOC_6H_4COOH \rightarrow$ .... [1]

b) Justify the industrial use of ethanoic anhydride rather than ethanoyl chloride in the manufacture of aspirin. [3]

....

....

....

....

....

4 An ester with the structural formula $CH_3CH_2COOCH_3$ can be produced by the reaction between propanoyl chloride and methanol.

a) Name the ester. .... [1]

b) Outline and name the mechanism for this reaction. [5]

Name of mechanism: ....

Benzene

🔒 Key words

❓ Key questions

What is benzene?

Why does benzene favour substitution reactions over addition reactions?

What happens in a Friedel–Crafts acylation?

✔ Summary

Aromatic chemistry

Benzene, C_6H_6, is:

- an unsaturated hydrocarbon
- a simple molecule with covalent bonds between the carbon atoms that are intermediate in length between single and double carbon bonds
- a non-polar and planar molecule
- more likely to undergo substitution reactions than addition reactions so, will not react with bromine water.
- toxic and banned from use in schools.

In benzene, the p electrons are delocalised above and below the plane of the atoms.

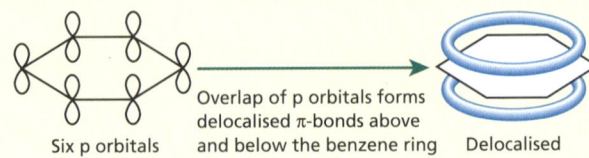

Six p orbitals   Overlap of p orbitals forms delocalised π-bonds above and below the benzene ring   Delocalised

Thermodynamic data of the hydrogenation of benzene compared to theoretical cyclohexa-1,3,5-triene shows that benzene is more stable than the theoretical molecule. This increased stability is known as the **delocalisation energy** and is due to the delocalised p electrons.

⬡ + 3H₂ → ⬡ $\Delta H^\ominus$ −208 KJ mol⁻¹

⬡ + H₂ → ⬡ $\Delta H^\ominus$ −120 KJ mol⁻¹

⬡ + 3H₂ → ⬡   Predicted $\Delta H^\ominus$ −360 KJ mol⁻¹

enthalpy

−360 KJ/mol Theoretical value

ΔH = −152 KJ/mol delocalisation energy

ΔH = −208 KJ/mol actual value

Types of reaction

Benzene doesn't undergo addition reactions as this would cause a loss of the stabilising effect of delocalised electrons. As benzene has a high electron density, it can undergo nucleophilic substitution reactions.

Nitration can be used in synthesis reactions to form explosives and amines. This involves an acid-base reaction between concentrated sulfuric and nitric acids to make the electrophile: $HNO_3 + 2H_2SO_4 \rightarrow NO_2^+ + 2HSO_4^- + H_3O^+$

This electrophile can then substitute a hydrogen atom on the benzene ring:

Sulfuric acid is a catalyst as the proton generated in the last step of the mechanism can combine with the HSO_4^- ion to regenerate the sulfuric acid.

Friedel–Crafts acylation reactions use a halogen carrier (a catalyst of $AlCl_3$) to react an acyl chloride with a benzene ring to form a ketone. This reaction proceeds at a moderate heat (50°C) under reflux.

Mechanism

The H⁺ ion reacts with the $AlCl_4^-$ to reform $AlCl_3$ catalyst and HCl.

$H^+ + AlCl_4^- \rightarrow AlCl_3 + HCl$

$RCOCl + AlCl_3 \rightarrow RCO^+ + AlCl_4^-$

Benzene

1. Which reagents are needed for the nitration of benzene? Tick **one** box. [1]

Concentrated sulfuric acid and concentrated nitric acid ☐

Dilute sulfuric acid and silver nitrate solution ☐

Concentrated sulfuric acid and aluminium chloride ☐

Nitric acid and silver nitrate solution ☐

2. What is the role of $AlCl_3$ in the acylation of benzene? Tick **one** box. [1]

Catalyst ☐   Electrophile ☐

Nucleophile ☐   Reagent ☐

3. Methanoyl chloride, HCOCl, can react with benzene to form benzaldehyde.

   a) Describe the reaction conditions. [3]

   b) Write a balanced equation for this reaction. [1]

   c) Name the mechanism for this reaction. [1]

4. The relative stability of benzene compared to cyclohexa-1,3,5-triene is shown.

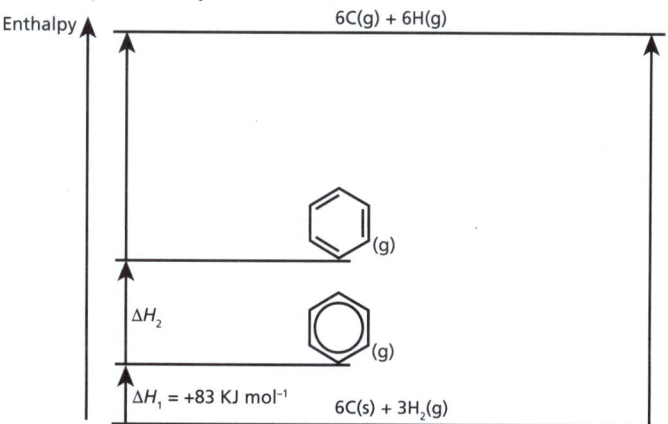

Thermodynamic data is shown in the table.

Enthalpy change	ΔH/kJ mol^{-1}
$\Delta H_{atom}^{\varnothing}(C)$	+715
$\Delta H_{atom}^{\varnothing}(H_2)$	+218
Mean bond enthalpy C–C	+348
Mean bond enthalpy C=C	+612
Mean bond enthalpy C–H	+412

   a) Use the thermodynamic data to calculate the delocalisation energy, ΔH_2. [3]

   b) Explain why benzene is more thermodynamically stable than the theoretical cyclohexa-1,3,5-triene. [1]

Amines

Key questions

What is an amine?

How do you make aliphatic amines?

How do you make aromatic amines?

What mechanism occurs in the reactions between ammonia and primary amines with acyl chlorides and acid anhydrides?

Types of amine

Amines are compounds based on ammonia where hydrogen atoms have been replaced by alkyl or aryl groups. Amines are weak bases, nucleophiles and ligands.

Aliphatic amines are organic substances and do not contain a six-carbon ring with delocalised electrons.

Primary aliphatic amines have the functional group R–NH$_2$ and can be prepared by:

- reacting excess ammonia with halogenoalkanes in a closed system; this reaction is a nucleophilic substitution reaction
- the reduction of nitriles.

Primary amine (1°)
1 alkyl group and 2 hydrogens

Secondary amine (2°)
2 alkyl groups and 1 hydrogen

Tertiary amine (3°)
3 alkyl groups

Quaternary ammonium ion
4 alkyl groups and a positive charge

Aromatic amines are used in the manufacture of dyes and made by the reduction of nitro compounds with tin and concentrated sulfuric acid.

Strength of amines

Strength is a measurement of the ionisation in aqueous solution. Primary aliphatic amines are stronger than ammonia, as the lone pair of electrons are more available. This means that the lone pair of electrons on the nitrogen is more available to accept protons in the primary aliphatic amine than in ammonia. Aromatic amines do not form basic solutions as the nitrogen lone pair of electrons can join the delocalised electrons on the aromatic ring, making it unavailable for reaction.

Nucleophilic substitution reactions

Ammonia can react with halogenoalkanes to produce a variety of products:

Step 1 to make primary amine	$NH_3 + RX \rightarrow [RNH_3]^+X^-$ $[RNH_3]^+X^- + NH_3 \rightarrow RNH_2 + [NH_4]^+X^-$
Step 2 to make secondary amine	$RNH_2 + RX \rightarrow [R_2NH_2]^+X^-$ $[R_2NH_2]^+X^- + NH_3 \rightarrow R_2NH + [NH_4]^+X^-$
Step 3 to make tertiary amine	$RNH + RX \rightarrow [R_3NH]^+X^-$ $[R_2NH]^+X^- + NH_3 \rightarrow R_3N + [NH_4]^+X^-$
Step 4 to make a quaternary ammonium salt (used as cationic surfactants)	$R_3N + NH_3 \rightarrow [R_4N]^+X^-$

Nucleophilic addition-elimination

Ammonia and primary amines react with acyl chlorides and acid anhydrides in a nucleophilic addition-elimination mechanism (see page 120).

Summary

Amines

(1) Which compound is the strongest base? Tick **one** box. [1]

Ammonia ☐

Ammonium chloride ☐

Phenylamine ☐

Cyclohexylamine ☐

(2) Which type of reaction is used to make the cationic surfactant $[(CH_3)_3N(CH_2)_{15}CH_3]Cl$ from a nitrile? Tick **one** box. [1]

Nucleophilic substitution ☐

Electrophilic substitution ☐

Nucleophilic addition-elimination ☐

Acid-base reaction ☐

(3) Methylbenzene can undergo a three-step reaction process to make 4-methyl-N-phenylacetamide.

a) Give the reagents needed in step 1. [2]

....

....

b) Name organic substance B. [1]

....

c) Name the mechanism for step 3. [2]

....

(4) An amine can be prepared by reacting excess ammonia with 2-bromopropane.

a) Outline a mechanism for this reaction. [4]

b) Compare the basic nature of the product of this reaction with ammonia. [4]

....

....

....

....

Condensation polymers

Key words

Key questions

What is a condensation polymer?

How are polyamides made?

How are polyesters made?

Why are polyalkenes not biodegradable but condensation polymers are?

Condensation polymers

Condensation polymers are formed when monomers react together to make a long-chain molecule (polymer) and a small molecule, e.g. water. They can be formed by:

- dicarboxylic acids and diols reacting together to make **polyesters**, e.g. Terylene

Benzene-1,4-dicarboxylic acid
Ethane-1,2-diol

The −1 here is because at each end of the chain the H and OH are still present

Terylene fabric is used in clothing, tire cords

- dicarboxylic acids and diamines reacting together to form **polyamides**, e.g. nylon 6,6 (used to make clothing, ropes, carpets, parachutes and other items requiring high strength and durability) Nylon 6,6 - a common polyamide and Kevlar (used in bullet-proof vests and composites in boat construction).

hexanedioic acid
Hexane-1,6-diamine

- amino acids reacting together to make **proteins**.

Intermolecular forces

There are induced dipole-dipole forces between all polymer chains. In addition:

- polyesters have permanent dipole forces as the C=O is polar
- polyamides and proteins have hydrogen bonds between the oxygen in the CO and the H in the amine group.

Used polymers

After being used, polymers can be:

- reused – reduces waste, conserves resources, reduces greenhouse gases
- recycled – reduces waste, conserves resources and uses less energy than making new polymers
- biodegraded – broken down in the environment
- combusted – produces greenhouse gas emissions and may cause air pollution, can be used to generate electricity
- buried in landfill – the least sustainable outcome because polyalkenes are inert due to the fact they contain many strong covalent bonds that will remain chemically unchanged in landfill for hundreds of years or more.

Polyesters and polyamides can be broken down by hydrolysis and are biodegradable. For polyesters, the reaction rate is quicker in basic conditions. For polyamides, it is quicker in acidic conditions.

Summary

Condensation polymers

(1) Which polymer **cannot** biodegrade? Tick **one** box. [1]

Nylon 6,6 ☐

Terylene ☐

Kevlar ☐

Poly(but-1-ene) ☐

(2) Which polymer does **not** contain hydrogen bonds between the polymer chains? Tick **one** box. [1]

Protein ☐

Nylon 6,6 ☐

Kevlar ☐

Terylene ☐

(3) Ethanedioic acid (HOOCCOOH) reacts with propane-1,3-diol (HOCH$_2$CH$_2$CH$_2$OH) to form a condensation polymer.

a) Name the type of condensation polymer formed. [1]

b) Draw the repeating unit of this polymer. [1]

(4) Nylon 4,6 can be formed from the polymerisation of five molecules of butane-1,4-diamine and five molecules of hexanedioic acid.

a) Complete the equation for this reaction by deducing the values of x and y. [2]

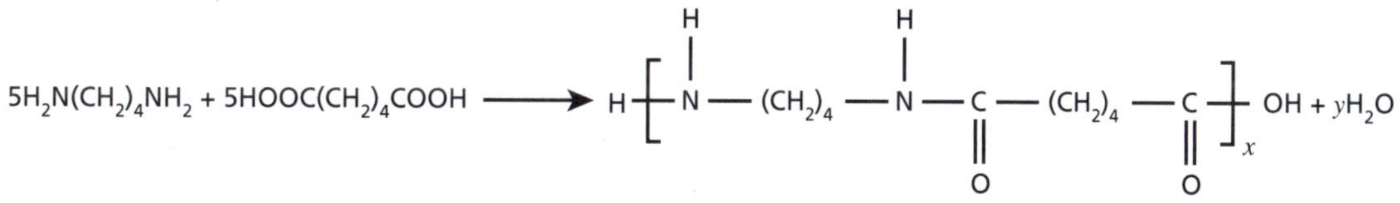

$$5H_2N(CH_2)_4NH_2 + 5HOOC(CH_2)_4COOH \longrightarrow H\left[\begin{array}{c} H \\ | \\ N \end{array} - (CH_2)_4 - \begin{array}{c} H \\ | \\ N \end{array} - \begin{array}{c} C \\ || \\ O \end{array} - (CH_2)_4 - \begin{array}{c} C \\ || \\ O \end{array}\right]_x OH + yH_2O$$

$x =$ ....

$y =$ ....

b) Name the intermolecular forces that occur between the polymer chains. [1]

c) Explain why nylon 4,6 is biodegradable. [2]

Organic chemistry **127**

Amino acids and proteins

🔑 Key words

❓ Key questions

What is an amino acid?

What is a zwitterion?

How can amino acids be identified?

How can mixtures of amino acids be separated and analysed?

Amino acids and the structure of proteins

Amino acids have two functional groups: a basic amine group (–NH₂) and an acidic carboxylic acid group (–COOH).

When pH changes, the structure of the amino acid can also change in an equilibrium system.

Low pH ⇌ Zwitterion ⇌ High pH

Amino acids are monomers that can be joined together by peptide links in sequences to form proteins. Proteins have these structures:

Primary structure – the linear sequence of amino acids	**Secondary structure** – the folding pattern of the polypeptide backbone due to the presence of hydrogen bonds, e.g. α-helix and β–pleated sheets.	**Tertiary structure** – the 3D shape of the protein held in place by S–S cross links, hydrogen bonds and sometimes by ionic interactions.
Formation of tripeptide chain — glycine; –2 H₂O; tripeptide chain	α-helix; β-pleated sheet	Hydrophobic interactions; Polypeptide backbone; Hydrogen bond; Disulfide bridge; Ionic bond

The peptide links can undergo hydrolysis, which breaks the polymer chain and re-forms the constituent amino acids.

Thin-layer chromatography (TLC)

Mixtures of amino acids can be separated and analysed using TLC. The amino acids are not initially visible and developing agents like ninhydrin (makes amino acids pink) or UV light (makes the amino acids fluoresce) are used to locate them on the TLC plate so the R_f value can be calculated. See Required Practical 12 on page 151 to review how to complete a TLC analysis.

$$R_f = \frac{\text{distance travelled by substance}}{\text{distance travelled by solvent}}$$

1. **Set up Chromatography paper**
2. **Analyse Chromatogram.** By comparing the unknown mixture of amino acids to the position of known amino acids on the chromatogram.

Original amino acid mixture has separated out into two different amino acids

This chromatogram shows that the amino acid mixture contains lysine and alanine but does not contain methionine

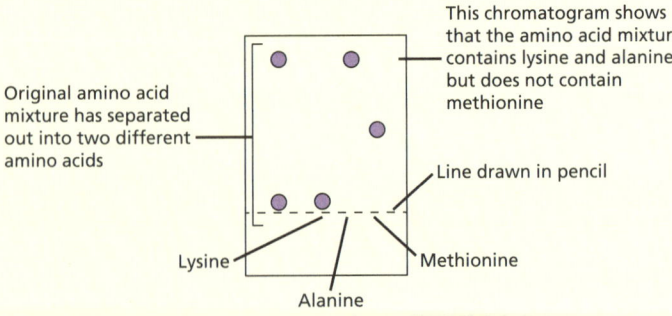

Line drawn in pencil

Lysine   Alanine   Methionine

✔ Summary

Amino acids and proteins

1 Which type of interaction between polypeptide chains is mainly responsible for forming an alpha helix? Tick **one** box. [1]

Hydrogen bonds ☐

Permanent dipole-dipole forces ☐

Induced dipole-dipole forces ☐

Covalent bonds ☐

2 Which diagram shows the structure of a zwitterion of an amino acid? Tick **one** box. [1]

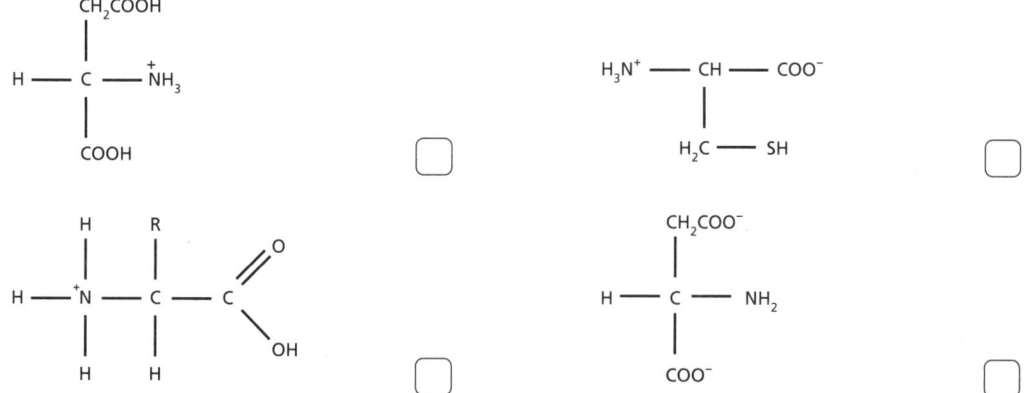

3 A student hydrolyses a sample of endomorphin-2 to break it down into its constituent amino acids. The resultant mixture of amino acids is analysed using TLC.

a) Explain why the amino acids separate on the TLC plate. [1]

b) i) Suggest a suitable developing agent to use in this investigation. [1]

ii) Why is a developing agent needed? [1]

c) The R_f value of glycine was 0.34

Predict the position of the glycine on a chromatogram where the solvent moved up the TLC plate by 80 mm. [2]

4 The simplified structure of an amino acid is shown below.

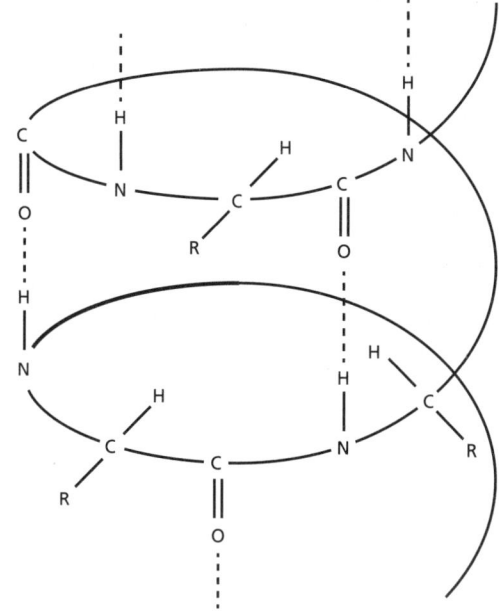

a) Name the protein structure shown. [1]

b) Explain the part of the structure shown by dotted lines. [4]

Enzymes

Key words

Key questions

What is an enzyme?

What is an enzyme inhibitor?

How can enzyme inhibitors be a drug?

How can new enzyme inhibitors be developed?

Enzymes are biological catalysts made from proteins

Enzymes are specific to certain substrates and can bind to the stereospecific active site.

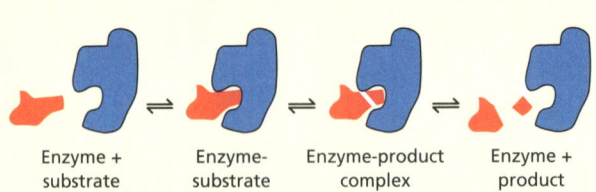

Enzyme + substrate ⇌ Enzyme-substrate complex ⇌ Enzyme-product complex ⇌ Enzyme + product

If the active site changes shape, it is said to be denatured and the enzyme can no longer bind with the substrate and affect the rate of reaction. Denaturing can occur with a pH change or an increase in temperature.

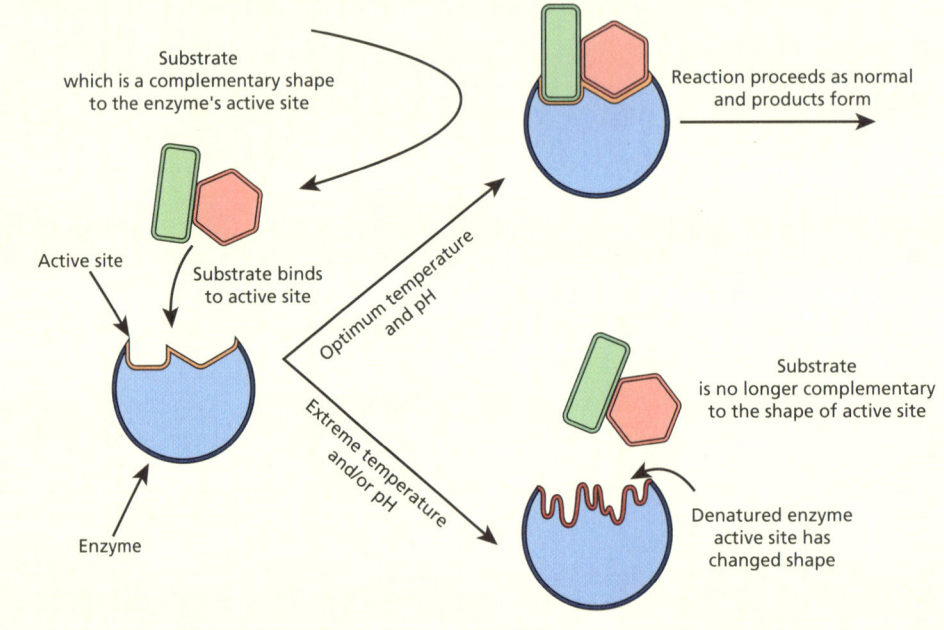

Enzyme inhibitors

Enzyme inhibitors are similar in shape to the substrate and fit into the active site. This prevents the enzyme from binding to the substrate and reduces the concentration of the active enzyme.

Competitive Inhibition

Enzyme + Substrate + Inhibitor → Enzyme - Inhibitor complex → Inhibitor binds to the active site. Directly blocks the active site. NO REACTION

Drugs change the way the body works. Medicines are drugs that are used to cure disease or treat symptoms. As the stereospecific active site can only bind to one enantiomer, computers are used to help with the challenging task of designing new drugs that are enzyme inhibitors.

Some medicines, like ibuprofen, are enzyme inhibitors.

Ibuprofen is a competitive inhibitor of the enzyme cyclooxygenase, helping to reduce the synthesis of biochemicals that cause inflammation and pain.

Ibuprofen

$C_{13}H_{18}O_2$

Summary

Enzymes

1 Which statement about enzymes is correct? Tick **one** box. [1]

Enzymes contain a stereospecific active site that can only bond to one enantiomeric form of a substrate ☐

Enzymes are made from amino acids which form an addition polymer chain ☐

A specific enzyme will operate in all pH levels ☐

Enzyme inhibitors increase the efficiency of enzymes and increase the rate of reaction further ☐

2 Which of the following molecules could be used as an enzyme inhibitor? Tick **one** box. [1]

$CH_3CH(CH_3)CH_2CH_3$ ☐

$CH_3 - CH - COOH$ with OH ☐

1, 3-dichloropropeane ☐

3 This question is about enzymes.

a) Define the term 'enzyme'. [2]

....

....

b) Explain why enzymes are specific to a particular substrate. [2]

....

....

....

c) Explain the effect of enzyme inhibitors on the action of enzymes. [3]

....

....

....

....

4 A racemic mixture of ibuprofen is made during the industrial manufacture of the anti-inflammatory.

Isomer F   **Isomer G**

Only isomer F is biologically active.

Suggest how an enzyme can be used to selectively remove isomer G using hydrolysis. [4]

....

....

....

....

....

....

DNA

Key questions

How is DNA a polymer?

What is the structure of DNA?

How does cisplatin work as an anti-cancer drug?

DNA (deoxyribonucleic acid)

DNA is a polymer in which two complementary strands form a double helix shape. The monomers are called **nucleotides**.

Each nucleotide consists of a phosphate ion, 2-deoxyribose (a pentose sugar) and one of the four bases (adenine, cytosine, guanine or thymine). All the structures that make up DNA are given in the exam data booklet (see page 220).

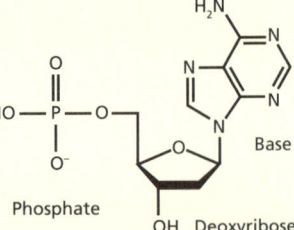

A nucleotide

Phosphate
Base
Deoxyribose

Structure and bonding in DNA

Covalent bonds link the nucleotides between the phosphate group of one nucleotide and the 2-deoxyribose of another nucleotide. The resulting polymer chain is a sugar-phosphate-sugar-phosphate polymer with bases attached to the sugars in the chain.

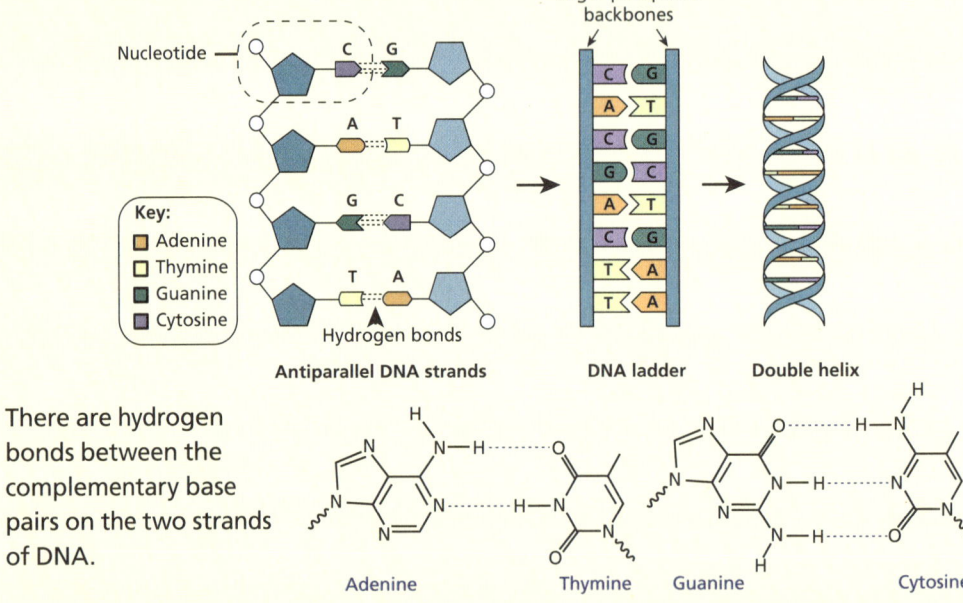

Nucleotide

Sugar-phosphate backbones

Key:
☐ Adenine
☐ Thymine
■ Guanine
■ Cytosine

Hydrogen bonds

Antiparallel DNA strands   **DNA ladder**   **Double helix**

There are hydrogen bonds between the complementary base pairs on the two strands of DNA.

Adenine   Thymine   Guanine   Cytosine

Cisplatin

Pt(II) complex cisplatin is a chemotherapy drug used in the treatment of cancer. The mechanism of the drug involves ligand replacement reaction. The N on two guanine bases on the DNA are exchanged for two Cl ligands on the cisplatin, causing the Pt to bind to the DNA and prevent replication of cells.

Cisplatin is not selective and this binding to biological molecules can happen to healthy cells. It is particularly noticeable in fast-replicating cells (e.g. the skin, hair and digestive system lining), resulting in side effects such as hair loss and digestive discomfort. Over time, the body can become resistant to the drug, making it less effective the more it is used. Whether the potential benefits outweigh the adverse effects is a key consideration in deciding to use a certain drug. Smaller quantities of a drug are often prescribed over a longer period to minimise the side effects and give time for the body to recover.

Cisplatin

Sugar phosphate backbone

Summary

DNA

1 Which pair of bases can link two strands of DNA with three hydrogen bonds? Tick **one** box. [1]

Cytosine and guanine ☐ Adenine and thymine ☐

Guanine and thymine ☐ Thymine and cytosine ☐

2 How does cisplatin bind to a DNA molecule? Tick **one** box. [1]

Replacement of one NH_3 ligand with the N on one guanine base ☐

Replacement of two Cl^- ligands with two N on two guanine bases ☐

Addition of ligands from two N bases ☐

Replacement of two ligands with two N from any base on the DNA strand ☐

3 One use of cisplatin, $[Pt(NH_3)_2Cl_2]$, is as an anti-cancer drug.

a) Explain the effect of cisplatin on cancer cells. [2]

b) A complex ion is formed after cisplatin enters a cell and one of the chloride ligands is replaced by a water molecule. Give the equation for this reaction. [2]

c) Complete the diagram to show how cisplatin binds to DNA. [2]

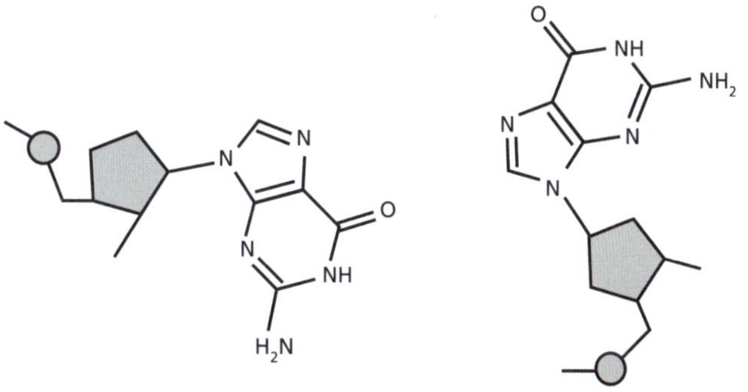

4 DNA is one of the molecules of life. The diagram below shows part of a DNA strand.

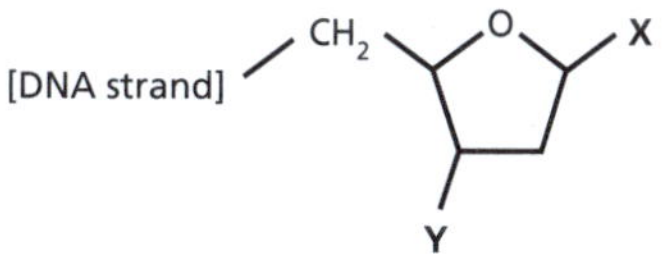

a) Name the groups labelled as X and Y. [2]

X = ... Y = ...

b) Describe the bonding in a single strand of DNA. [3]

c) Describe the structure of DNA. [3]

Organic synthesis

Key words

Key questions

What is organic synthesis?

Why do process chemists try to minimise the number of steps in a synthesis reaction?

Why do process chemists try not to use solvents in a synthesis reaction?

How do process chemists try to make synthesis reactions sustainable?

Organic synthesis

Process chemists design synthesis reactions that aim to:

- have a high yield and a high atom economy
- achieve a fast rate of production
- minimise energy consumption
- avoid using solvent
- use the least hazardous materials possible
- have as few steps as possible

Examples of reactions used for synthesis

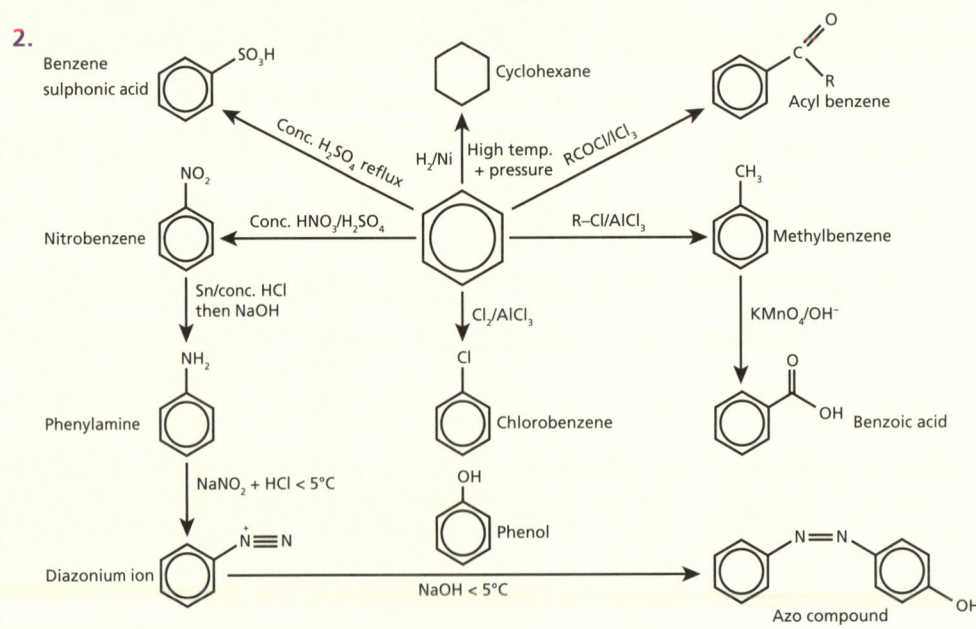

* Makes the amine with one more carbon atom: propylamine not ethylamine as shown in diagram.

✔ Summary

Organic synthesis

(1) What is the name of compound X in the synthetic route shown below? Tick **one** box. [1]

Compound **X** $\xrightarrow[\text{H}_2\text{SO}_4]{\text{Concentrated}}$ Intermediate $\xrightarrow[\text{H}_2\text{O}]{\text{Excess}}$ Alcohol

But-1-ene ☐ Butanal ☐ Butanoic acid ☐ Butane ☐

(2) What are the reagents for step 1 in the synthetic route shown below? Tick **one** box. [1]

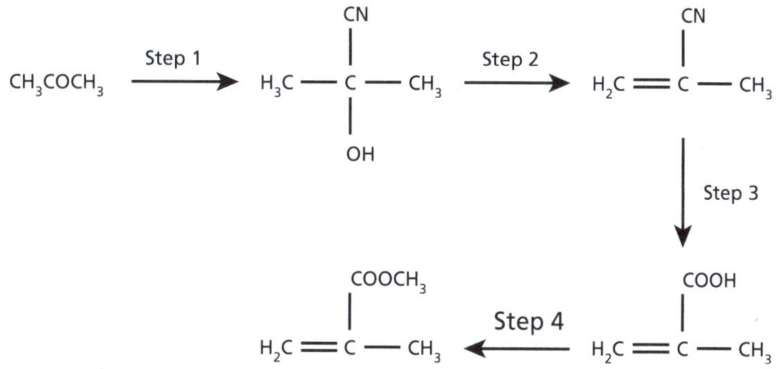

NaCN then dilute HCl ☐ Warm aqueous H_2SO_4 ☐
Warm ethanoic H_2SO_4 ☐ Hot ethanolic KOH ☐

(3) The diagram below shows the synthetic route to make a diester.

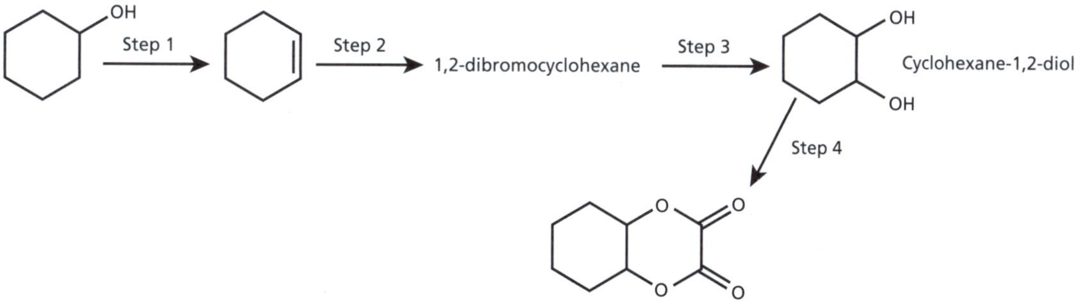

a) Give the reagents needed for steps 1, 2 and 3. [3]

Step 1: _____

Step 2: _____

Step 3: _____

b) Suggest why chemists usually aim to design production methods with few steps and high atom economy. [2]

(4) A synthetic route is shown below.

Ethene $\xrightarrow{\text{Step 1}}$ ClCH$_2$CH$_2$Cl $\xrightarrow{\text{Step 2}}$ C$_4$H$_4$N$_2$ $\xrightarrow{\text{Step 3}}$ H$_2$N(CH$_2$)$_4$NH$_2$

a) Name the chemical reactions that occur in steps 1 and 2. [2]

Step 1: _____ Step 2: _____

b) Suggest the reagents needed for each step. [3]

Step 1: _____

Step 2: _____

Step 3: _____

Nuclear magnetic resonance

Key words

Key questions

How are the structures of new compounds confirmed?

Nuclear magnetic resonance (NMR)

NMR gives information on the position of ^{13}C and ^{1}H in an organic substance.

The organic substance is dissolved in a solvent that doesn't contain protons, e.g. CCl_4, and is then put in a magnetic field. Some nuclei are affected by the magnetic field and they align with it. Electromagnetic waves (radio waves) are then used to force the nuclei out of alignment and they resonate between these two positions.

The energy needed for resonance differs according to the environment that the atom is in. The recorded energy is compared to the energy required to make carbon or hydrogen resonate in tetramethylsilane (TMS). This measurement (in ppm), called **chemical shift**, uses the δ scale.

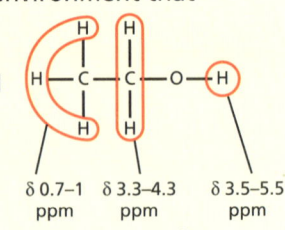

δ 0.7–1 ppm   δ 3.3–4.3 ppm   δ 3.5–5.5 ppm

In ethanol, the proton attached to the electronegative oxygen is very different to the protons in TMS and so has a high chemical shift. However, the protons in the methyl group are similar to the protons in TMS, so they have less of a chemical shift.

A computer produces a spectrum that can be interpreted to provide information about the structure.

Why is TMS used in NMR?

High resolution proton NMR

For high resolution proton NMR, the spectrum appears to have multiple peaks for a single signal. This is because of the spin–spin splitting pattern of adjacent, non-equivalent protons. To determine the splitting pattern, use the $n + 1$ rule.

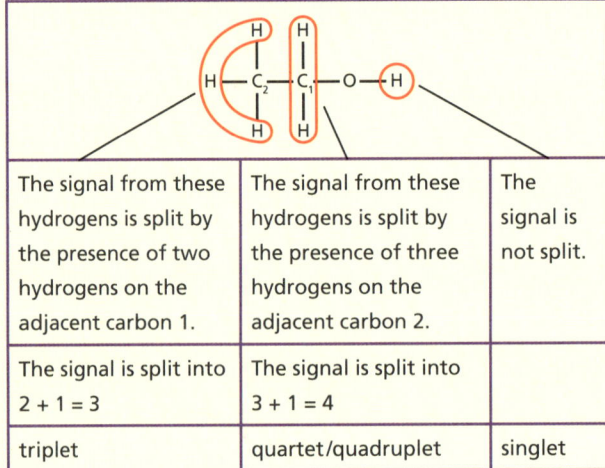

The signal from these hydrogens is split by the presence of two hydrogens on the adjacent carbon 1.	The signal from these hydrogens is split by the presence of three hydrogens on the adjacent carbon 2.	The signal is not split.
The signal is split into $2 + 1 = 3$	The signal is split into $3 + 1 = 4$	
triplet	quartet/quadruplet	singlet

What can NMR tell you?

What is the $n + 1$ rule?

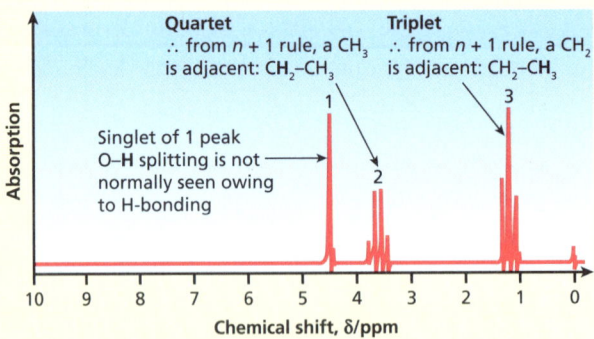

Proton NMR of ethanol

By comparing the chemical shift from spectra to data tables in the exam data booklet (as seen on page 219), you can suggest possible structures of molecules.

✔ Summary

Nuclear magnetic resonance

1. Which of the following is a suitable solvent for use in NMR? Tick **one** box. [1]

   Tetramethyl silane ☐   Tetrachloromethane ☐

   Water ☐   Ethanol ☐

2. Which of the following statements are **not** correct for the organic molecule shown below? Tick **one** box. [1]

   $$H_3C \!-\! C \!-\! O \!-\! CH_2 \!-\! CH_3$$
   $$\parallel$$
   $$O$$

   Its ^1H NMR spectrum has three peaks with an integration ratio of 2 : 3 : 3 ☐

   Its ^1H NMR spectrum has two peaks with an integration ratio of 2 : 3 ☐

   TMS is a suitable internal standard ☐

   Its ^{12}C NMR has higher chemical shift values than the 1H NMR ☐

3. Propan-2-ol is an organic substance. TMS was used as an internal standard when completing the ^{13}C NMR of propan-2-ol.

   a) Give **two** reasons why TMS is a suitable internal standard. [2]

   _____

   _____

   b) On the axes below, sketch the ^{13}C NMR spectrum for propan-2-ol. [2]

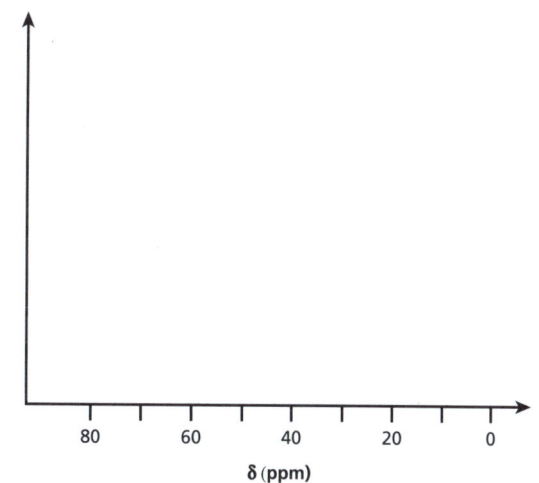

   δ (ppm)

4. An organic compound has the molecular formula $C_4H_8O_2$. IR spectroscopy shows the presence of an −OH from an alcohol group and a C=O functional group. A further test tube reaction shows the carbonyl group is a ketone.
   The table below shows the information from the ^1H NMR of the organic substance.

δ (ppm)	2.20	2.69	3.40	3.84
Integration value	3	2	1	1
Splitting pattern	Singlet	Triplet	Singlet	Triplet

   Use the information to determine the structure of the organic compound. [4]

Chromatography

Key words

Key questions

What can chromatography be used for?

How does chromatography separate mixtures?

What are the three types of chromatography?

What is retention time?

Chromatography: separating a mixture into its components

In chromatography, a complex mixture can be separated as it travels in a fluid solvent called the mobile phase, which passes through a fixed stationary phase. Different substances within the mixture travel at different speeds, so some move further than others in a given time.

Types of chromatography	
Thin-layer chromatography (TLC) A plate is coated with a solid and a solvent moves up the plate.	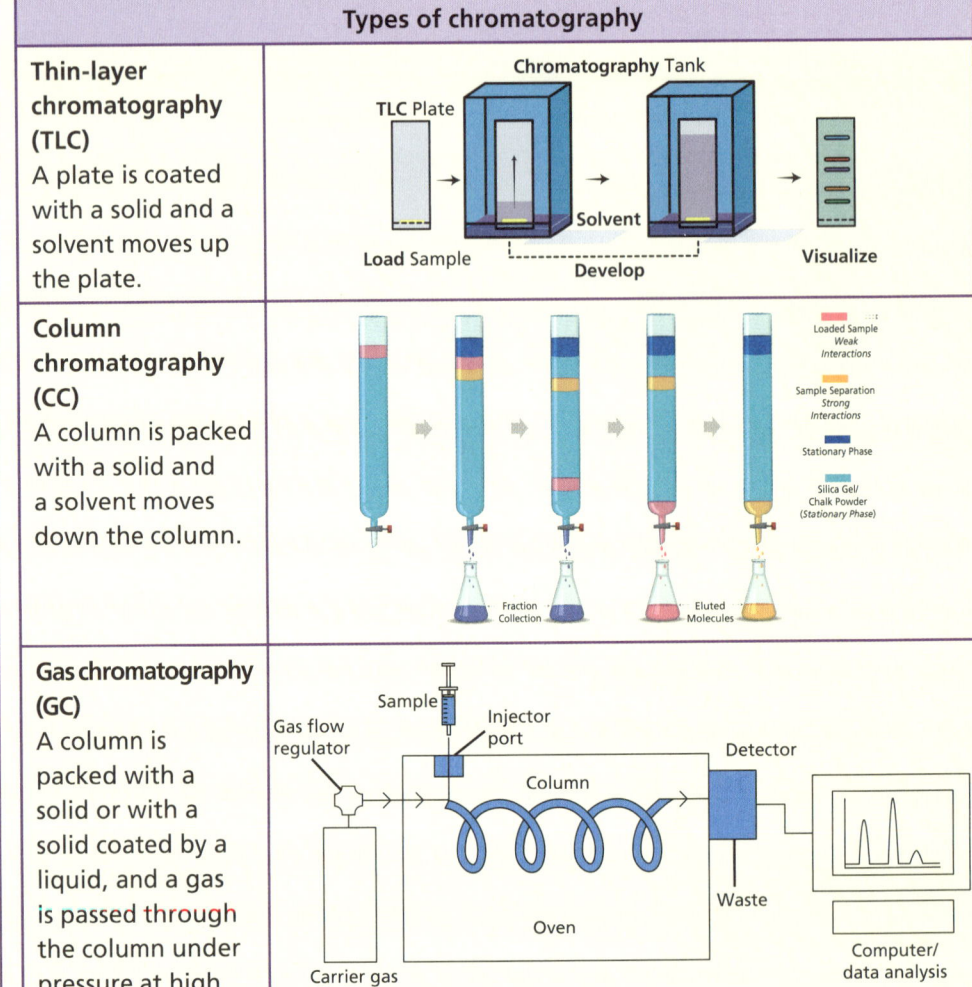
Column chromatography (CC) A column is packed with a solid and a solvent moves down the column.	
Gas chromatography (GC) A column is packed with a solid or with a solid coated by a liquid, and a gas is passed through the column under pressure at high temperature.	

Identifying substances

The retention time is the time taken for a fraction of the mixture to leave the chromatography column. The retention factor, R_f, is the ratio of how far the mixture travels compared to the solvent. This can be calculated for a substance and compared to known data to identify it.

$$R_f = \frac{\text{distance moved by spot}}{\text{distance moved by solvent front}}$$

GC-MS is gas chromatography linked to a mass spectrometer. These two machines allow the different fractions of a mixture to be instantly analysed.

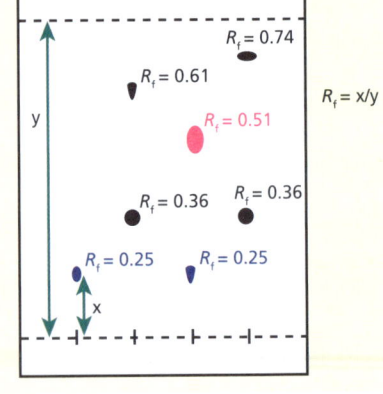

Summary

Chromatography

(1) What causes the separation in chromatography? Tick **one** box. [1]

The balance between solubility in the mobile phase and retention by the stationary phase ☐

The balance between solubility in the stationary phase and retention by the mobile phase ☐

The solubility of the substance in the stationary and mobile phases ☐

The reactivity of the substance with the mobile phase ☐

(2) Which type of chromatography is often used with mass spectrometry? Tick **one** box. [1]

Gas chromatography ☐

Paper chromatography ☐

Thin-layer chromatography ☐

Column chromatography ☐

(3) A sample of cyclohexanol has been contaminated with cyclohexene. The cyclohexene can be separated from the cyclohexanol by column chromatography using silica gel and hexane.

a) Identify the mobile and stationary phase in the column chromatography. [1]

Mobile phase = ...

Stationary phase = ...

b) Explain why cyclohexanol has a longer retention time than cyclohexene. [2]

....

....

....

(4) Using hydrolysis, a student produced a mixture of amino acids from endomorphin-2. The diagram shows the thin-layer chromatography (TLC) plate from the analysis of the mixture.

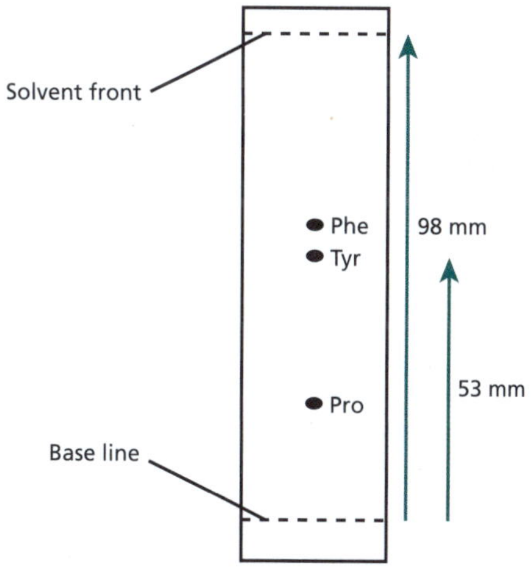

a) Suggest a suitable developing agent to visualise the proteins on the TLC plate.

.... [1]

b) Calculate the R_f value for Tyr. [2]

Required practical 1: Titration

Aim

To make up a volumetric solution and carry out a simple acid–base titration.

My lab notes

Making a standard solution

1. Use weighing by difference and a minimum of a 2 decimal place top pan balance to accurately measure the mass of the solid being used.
2. Dissolve the solid in a small volume of deionised water (must be less than 250 cm³).
3. Add the solution into a volumetric flask with washings and add water until the bottom of the meniscus is in line with the 250 cm³ graduation mark.
4. Put in the stopper and invert three times to form a homogenous solution.
5. Calculate the accurate concentration of the solution by:

$$\text{concentration (mol dm}^{-3}) = \frac{\text{mass (g)} \times 1000}{Mr \times 250}$$

6. Put a label on the volumetric flask with your name, the date, the name of the substance, the formula of the substance, the concentration of the substance and any hazards.

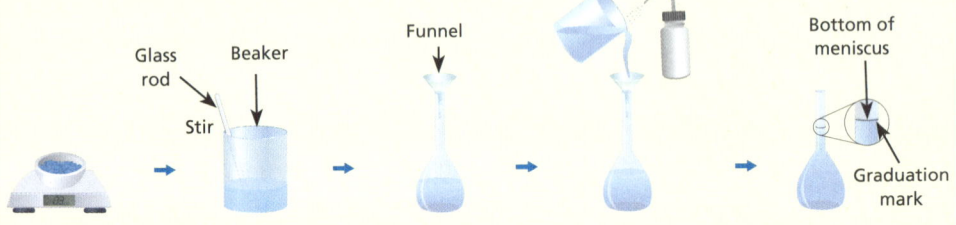

Glass rod   Beaker   Funnel   Bottom of meniscus

Stir   Graduation mark

Acid–base titration

1. Use clean dry glassware.
2. Put a waste beaker under the burette and use a funnel to carefully fill the burette with acid, ensuring that the filling is not above head height. Open the tap and make sure that the jet is full of solution. Turn off the tap and make a note of the starting volume.
3. Use a pipette filler to transfer exactly 25.00 cm³ of alkali into a conical flask and add a few drops of a suitable acid–base indicator (e.g. phenolphthalein) so that the colour is just observable.
4. For the rough titration, put the conical flask on a white tile under the burette and run in the acid while swirling the mixture. As soon as there is a persistent colour change, stop adding the solution and note the final volume. Calculate the rough titre.
5. For the accurate titrations, complete steps 2 and 3, then put the conical flask on the white tile under the burette. Quickly add 5 cm³ below the rough titration titre. Then add the solution drop wise, swirling continuously, as the colour change becomes more persistent. Reduce the flow rate and use the wash bottle to wash in any unreacted solution. When the colour change persists for 10 seconds, stop the titration and note the final volume. Calculate the titre and repeat until two concordant (within 0.1 cm³) titres are obtained.
6. Calculate the mean titre.

Pipette filler

Pipette filled with alkali solution

Burette filled with acid (unknown concentration)

Alkali solution

Alkali solution with suitable indicator

Analysis

Break down the calculation:

1. Write down a balanced equation for the reaction to determine the ratio of moles of acid to alkali involved.
2. Calculate the number of moles in the solution of known volume and concentration. You can work out the number of moles in the other solution from the balanced equation.
3. Calculate the concentration of the other solution.

	Rough titration	Accurate titration 1	Accurate titration 2
Final burette reading / cm³	37.60	36.20	38.40
Initial burette reading / cm³	1.80	0.00	2.10
Volume of acid used / cm³	35.80	36.20	36.30

Required practical 2: Enthalpy change

Aim

To measure the enthalpy change of a reaction.

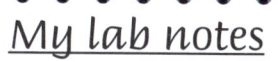

My lab notes

Example method: Measuring enthalpy change of a combustion reaction

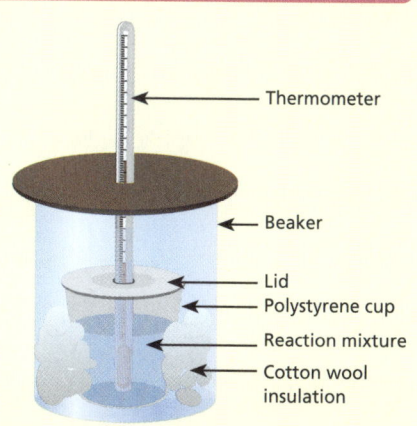

① Use a top pan balance to accurately measure the mass of a spirit burner with the fuel.

② Measure 100 g of water into a copper calorimeter and clamp in place above the spirit burner. You can reduce the heat loss to the surroundings (main error) by adding a lid and draft shield.

③ Measure the temperature of the water.

④ Light the fuel, stir the water and allow the temperature of the water to increase by at least 20°C. Record the highest temperature then extinguish the flame and allow the spirit burner to cool.

⑤ Re-weigh the spirit burner and calculate the mass of fuel used. Use this data to calculate the number of moles of fuel used.

Example method: Measuring enthalpy change of a reaction involving solutions

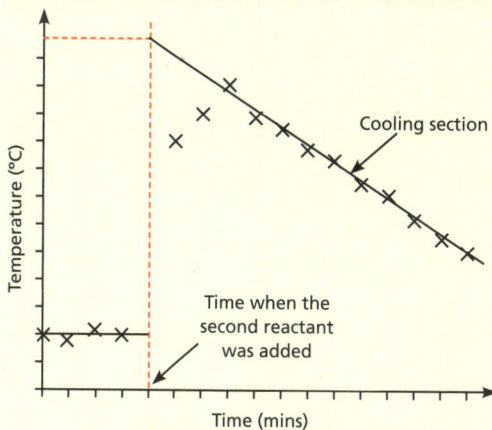

① Using a measuring cylinder, place 25 cm³ of the known concentration of a solution into a polystyrene cup or insulated beaker. Insulation is vital for reducing heat loss to the surroundings.

② Measure the temperature of the solution every 30 seconds for at least $2\frac{1}{2}$ minutes until the temperature is stable.

③ At 3 minutes, add excess solid reactant and stir.

④ Continue to stir and take the temperature every 30 seconds from $3\frac{1}{2}$ minutes until a total of 10 minutes has passed.

Analysis

① The maximum temperature change must be calculated:
   - For combustion calorimetry, this is calculated by $\Delta T = \text{final } T - \text{initial } T$
   - For coffee cup calorimetry, plot a graph and infer the maximum temperature change.

② Use the equation $Q = mC\Delta T$ where Q is the energy transferred (in J), m is the mass of the water/solution (in g), C is specific heat capacity (which is assumed to be for pure water at $4.18\,\text{J}\,\text{g}^{-1}\text{K}^{-1}$), and ΔT is the change in temperature (in K).

③ To determine an enthalpy change ΔH, you need to find the energy transferred per mole of the limiting reactant. Therefore, calculate the number of moles of the limiting reagent used. Then calculate the energy transfer for 1 mole of the limiting reactant. If the temperature increased, the ΔH will be negative (exothermic reaction). If the temperature decreased, the ΔH will be positive (endothermic reaction).

Required practical 3: Rate of reaction with temperature changes

Aim

To investigate how the rate of a reaction changes with temperature.

My lab notes

Example method: Measuring rate of reaction using the disappearing cross method

① Accurately measure $50\,cm^3$ of sodium thiosulfate solution into a conical flask.

② Place the conical flask on a piece of paper with a black cross drawn on it.

③ Measure $10\,cm^3$ of dilute hydrochloric into a measuring cylinder.

④ Put the solutions in a water bath and allow to reach the temperature of the water bath.

⑤ Add the hydrochloric acid to the conical flask and immediately start the stopwatch.

⑥ Continue to swirl the mixture and record the time when the cross is no longer visible when looking down through the solution.

⑦ Repeat the experiment and take a mean for each temperature and investigate different temperatures.

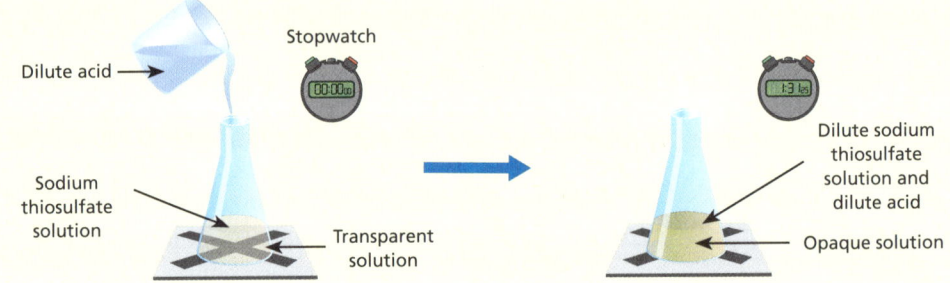

$$Na_2S_2O_3(aq) + 2HCl(aq) \longrightarrow 2NaCl(aq) + H_2O(l) + SO_2(g) + S(s)$$

Analysis

In this investigation:

- the concentration of the reactants are controlled variables
- the temperature is the independent variable
- the dependent variable is the time taken for the sulfur precipitate to cause a turbid solution where the cross is no longer visible.

A simple graph of the initial rate of reaction is shown.

The data should be processed to show that the initial rate of reaction is proportional to $\frac{1}{\text{time}}$ when the production of sulfur is assumed to be constant and a fixed amount.

The steps are:

① Calculate the mean temperature of each reaction mixture.

② Calculate $(\frac{1}{t})$ where t is the time taken for the cross to be obscured.

③ Plot a graph of $\frac{1}{t}$ on the y-axis against average temperature.

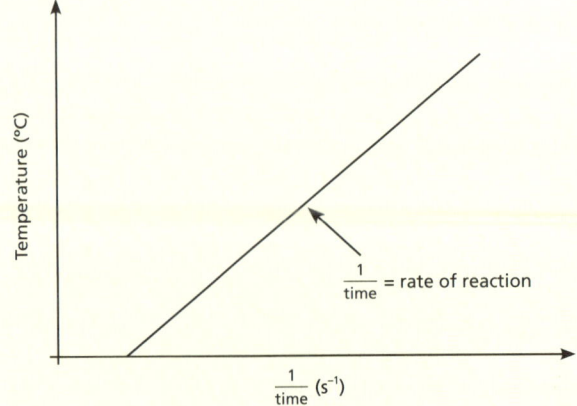

Required practical 4: Simple test-tube reactions

Aim

To carry out simple test-tube reactions to identify cations and anions.

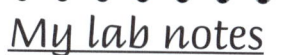

My lab notes

Testing for anions and cations

Test for	Example methods	Expected results
Group 2 cations	① Add 10 drops of unknown solution to a clean test tube. ② Add sodium hydroxide solution dropwise, shaking on each addition. Note observations and continue until NaOH is in excess.	All group 2 metal ions and Al^{3+} produce white precipitates and Mg^{2+} has no observable change.
	① Add 10 drops of unknown solution to a clean test tube. ② Add sulfuric acid solution dropwise, shaking on each addition. Note observations and continue until H_2SO_4 is in excess.	Ba^{2+} and Sr^{2+} ions produce white precipitates, and Mg^{2+} and Ca^{2+} ions produce a slightly white precipitate (for Mg^{2+} a colourless solution is observed on addition of excess H_2SO_4).
Ammonium cations	① Add 10 drops of unknown solution to a clean test tube. ② Add the same volume of NaOH solution, shake and warm in a water bath. ③ Use damp blue litmus paper to test fumes.	If NH_4^+ ions are present then $NH_3(g)$ will be produced, turning the blue litmus paper red and then white (bleach).

Ammonia gas released

Unknown substance and sodium hydroxide

Red litmus paper

Test for	Example methods	Expected results
Halide anions	① Add 10 drops of unknown solution to a clean test tube. ② Add 5 drops of dilute HNO_3 and 10 drops of $AgNO_3$ and shake. ③ Observe any precipitate.	Cl^- ions produce a white precipitate that appears to dissolve in dilute and concentrated NH_3. Br^- ions produce a cream precipitate that appears to dissolve in concentrated NH_3. I^- ions produce a yellow precipitate that doesn't appear to dissolve in any concentration of NH_3.

Add silver nitrate solution

Unknown halide solution dissolved in dilute nitric acid

Precipitate of silver halide formed

④ Add 10 drops of dilute NH_3 and shake. Observe.
⑤ Add 10 drops of concentrated NH_3 and shake. Observe.

Test for	Example methods	Expected results
Sulfate anions	① Add 1 cm^3 of unknown solution to a clean test tube. ② Add 1 cm^3 of dilute HCl and 1 cm^3 $BaCl_2$. ③ Observe any precipitate.	SO_4^{2-} ions produce a white precipitate.
Carbonate anions	① Add 3 cm^3 of unknown solution to a clean test tube. ② Add 3 cm^3 of dilute Na_2CO_3. ③ Observe and test any gas produced with limewater, $Ca(OH)_2(aq)$.	CO_3^{2-} ions produce effervescence and the gas causes limewater to turn cloudy.

Delivery tube

Bubbles of CO_2 gas

Hydrochloric acid

Limewater

Calcium carbonate

Test for	Example methods	Expected results
Hydroxide anions	① Add 1 cm^3 of unknown solution to a clean test tube. ② Add an acid-based indicator, such as litmus or universal indicator. Alternatively use a pH probe. ③ Observe the colour or read off the number from the pH probe.	Litmus will turn red and universal indicator will turn dark green, blue or purple. The pH probe will have a number greater than 7.0.

Analysis

You do not need to perform every test-tube reaction on every sample. As soon as you have determined the anion and the cation in the sample, there is no need to continue with any of the other tests.

Required practical 5: Distillation

Aim

To carry out the distillation of a product from a reaction.

My lab notes

Example method: Making a sample of a pure dry organic product

① Use a top pan balance to weigh by difference and determine the mass of cyclohexanol being used.

② Add 20 cm³ of cyclohexanol and approximately 8 cm³ of phosphoric acid with a few anti-bumping granules (to give a smooth boil) to a 50 cm³ flask.

③ Set up the distillation apparatus as shown in the diagram and gently heat the reaction mixture.

④ Collect all the distillate produced at a temperature below 100°C.

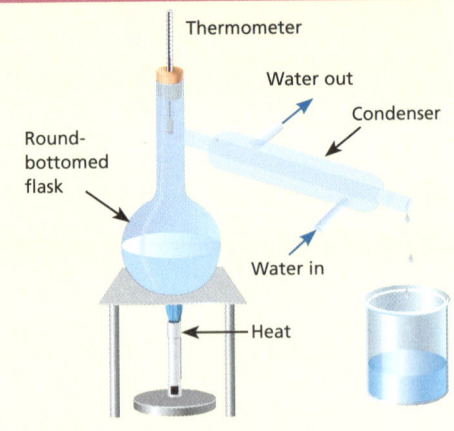

⑤ Add the distillate into a separating funnel and add 50 cm³ of saturated sodium chloride solution (which helps to separate the organic layer). Shake and collect the upper, organic layer.

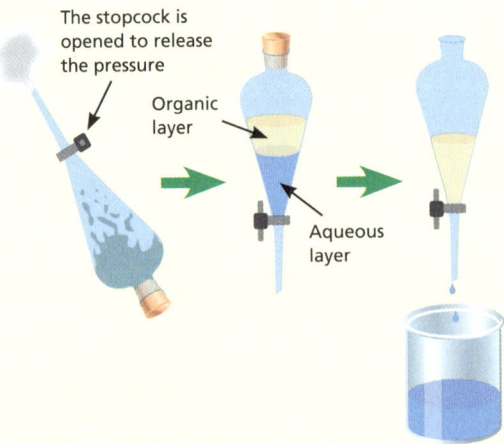

⑥ Add a drying agent (anhydrous calcium carbonate) until no more clumping. Allow to settle and decant off the colourless liquid into a clean, dry sample bottle.

⑦ Label the sample container with your name, the date, the substance, mass, and any hazards.

Expected results

Cyclohexanol will undergo a dehydration reaction catalysed by the phosphoric acid. Pure cyclohexene is the organic product that is collected.

Analysis

① Calculate the theoretical yield of cyclohexene.

Calculate the number of moles of cyclohexanol by

$$n = \frac{m}{M_r}$$

where n = amount (mol), M = mass (g) and M_r = relative formula mass = 100

The balanced symbol equation is: $C_6H_{11}OH \longrightarrow C_6H_{10} + H_2O$

This shows that there is a 1 : 1 ratio, so the amount of cyclohexanol = amount of cyclohexene.

Calculate the mass of cyclohexene using $n = \frac{m}{M_r}$ but the M_r of cyclohexene should be used (82).

② Calculate the percentage yield of your product.

$$\text{Percentage yield (\%)} = \frac{\text{actual yield (g)}}{\text{theoretical yield (g)}} \times 100$$

Required practical 6: Organic tests

Aim

To carry out simple test tube tests for alcohols, aldehydes, alkenes and carboxylic acids.

My lab notes

Testing for alcohols, aldehydes, alkenes and carboxylic acids

Test for	Example methods	Expected results
Alcohols, –OH	**Reactive metals** ① Add a small piece of sodium to a sample. ② Observe and test the gas with a lighted splint. Wood splint, Alcohol, Hydrogen gas, Sodium	Alcohols effervesce and the gas makes a squeaky pop noise with a lighted splint.
	Potassium dichromate, $K_2Cr_2O_7$ Add acidified potassium dichromate to the solution. Aqueous solution of acidified potassium dichromate(VI) Result if the chemical being tested is not a reducing agent Result if the chemical being tested is a reducing agent Acidified potassium dichromate(VI) remains orange — Sample of chemical being tested — Orange acidified potassium dichromate(VI) changes to green	With primary and secondary alcohols, the orange solution turns green. There is no observable change with tertiary alcohols.
Aldehydes, CHO	**Fehling's solution** ① To a clean test tube, add 5 drops of Fehling's solution and 5 drops of the unknown solution. ② Gently warm in a water bath for a few minutes. Negative test result (Ketone) Positive test result (Aldehydes)	A brick red precipitate is formed if an aldehyde is present.
	Tollens' reagent ① To a clean test tube, add 5 drops of Tollens' reagent and 5 drops of the unknown solution. ② Observe. Before After	A silver metal precipitate is formed (silver mirror) if an aldehyde is present.
	Potassium dichromate, $K_2Cr_2O_7$ Add acidified potassium dichromate to the solution.	Aldehydes change the orange solution to green.
Alkenes, C=C	① To a clean test tube, add 5 drops of bromine water and 5 drops of the unknown solution. ② Shake and observe. Bromine water (orange coloured) Saturated (no double bonds) Unsaturated (double bonds)	Alkenes decolourise the bromine water.
Carboxylic acids, –COOH	**Sodium hydrogen carbonate, $NaHCO_3$** ① Add 5 drops of sodium hydrogen carbonate solution and 5 drops of the unknown solution to a clean test tube. ② Observe and test any gas produced with limewater, $Ca(OH)_2$ (aq). Delivery tube Bubbles of CO_2 gas Hydrochloric acid Limewater Calcium carbonate	Carboxylic acids produce effervescence and the limewater changes from colourless to cloudy, confirming the presence of carbon dioxide.
	Measuring pH ① Add universal indicator or a pH probe. ② Determine the pH.	For a carboxylic acid, the pH will be less than 7.0.

Analysis

You do not need to ensure an uncontaminated sample is used for each chemical test to prevent false positive results. You do need to complete all the tests, as one organic substance could have more than one functional group.

Required practical 7: Rate of reaction

Aim

To measure the rate of reaction by an initial rate method and by a continuous monitoring method.

My lab notes

Example method: Initial rate, iodine clock reaction

① Add 25 cm³ of sulfuric acid, 20 cm³ of deionised water, about 1 cm³ of starch solution, 5.0 cm³ of potassium iodide solution and 5.0 cm³ of sodium thiosulfate solution to a clean, dry 250 cm³ beaker.

② Stir the mixture and add 10.0 cm³ hydrogen peroxide solution and immediately start the timer.

③ Continuously stir the mixture and stop the timer when the mixture turns blue-black.

④ Repeat so that five different concentrations of potassium iodide are used.

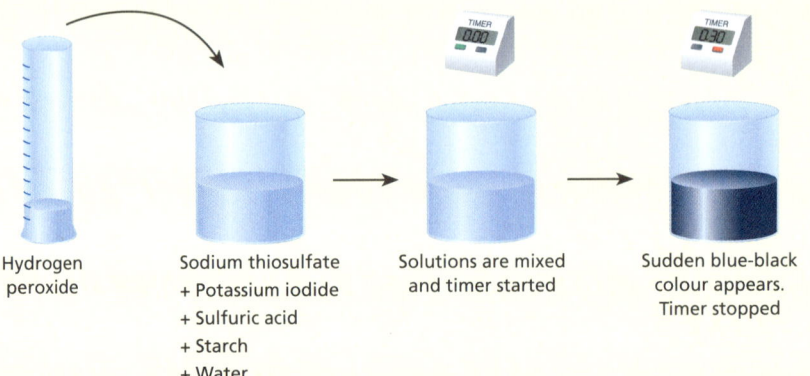

Hydrogen peroxide

Sodium thiosulfate
+ Potassium iodide
+ Sulfuric acid
+ Starch
+ Water

Solutions are mixed and timer started

Sudden blue-black colour appears. Timer stopped

Analysis

Plot a graph of initial rate (y-axis) against concentration (x-axis) to determine the order with respect to iodide ions in acidic solution.

Example method: Continuous method

① Add 50 cm³ of hydrochloric acid with a 6 cm strip of magnesium ribbon to a conical flask.

② Quickly attach the bung and gas syringe, then start the timer.

③ Record the volume of hydrogen gas collected every 10 seconds for 3 minutes.

④ Repeat steps 1 to 3 using different concentrations of acid.

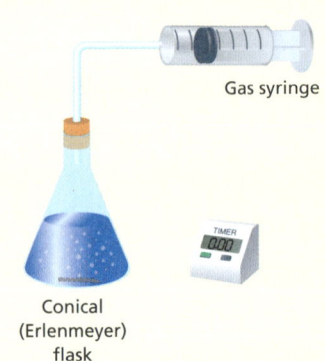

Gas syringe

Conical (Erlenmeyer) flask

Analysis

Plot a graph of volume of hydrogen produced (y-axis) against time in seconds (x-axis) for each hydrochloric acid concentration and draw the five lines of best fit. Calculate the gradient of a tangent to each line at $t = 0$ s to determine the initial rate of reaction for each concentration.

Use these data to determine the effect on the rate of reaction of changing the concentration of the acid.

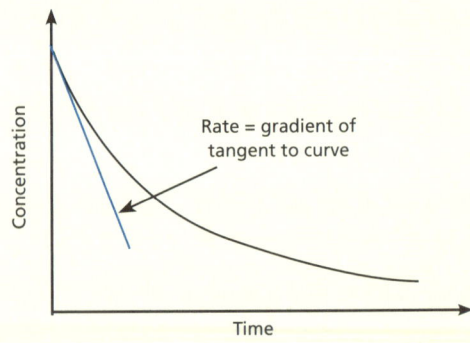

Concentration

Rate = gradient of tangent to curve

Time

Required practical 8: Electrochemical cells

Aim

To measure the electromotive force (EMF) of an electrochemical cell.

My lab notes

Example method

① Use sandpaper and propanone to clean the surface of the metals.

② To make a half-cell, half fill a beaker with 1.00 mol dm^{-3} of a metal ion solution. Put in a clean electrode made from the metal that is in the compound.

③ Soak a piece of absorbent paper into saturated brine and put one end in the solution in one beaker and the other end in the solution of the other beaker.

④ Use crocodile clips and wires to connect the voltmeter between the two metals.

⑤ Record the reading on the voltmeter.

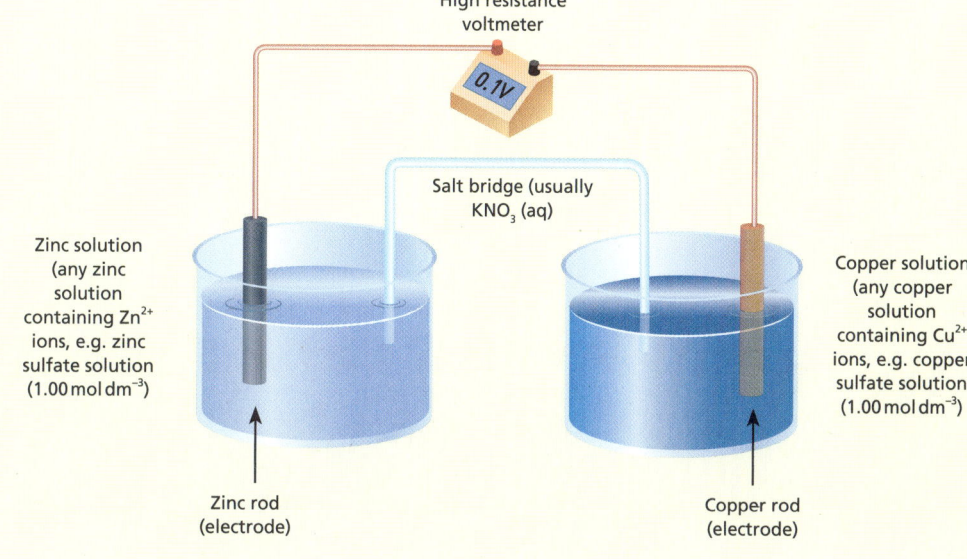

High resistance voltmeter

0.1V

Salt bridge (usually KNO$_3$ (aq))

Zinc solution (any zinc solution containing Zn^{2+} ions, e.g. zinc sulfate solution (1.00 mol dm^{-3}))

Copper solution (any copper solution containing Cu^{2+} ions, e.g. copper sulfate solution (1.00 mol dm^{-3}))

Zinc rod (electrode)

Copper rod (electrode)

$$Zn(s) \mid Zn^{2+}(aq) \parallel Cu^{2+}(aq) \mid Cu(s)$$

Analysis

Use different combinations of half-cells and record the potential difference, measured on the voltmeter. This is the experimental EMF of the electrochemical cell. The higher the EMF, the larger the difference in reactivity between the metals.

Experimental results are usually quite different to theoretical results as standard conditions are not maintained.

Required practical 9: Titration curves

Aim

To investigate how pH changes when a weak acid reacts with a strong base and when a strong acid reacts with a weak base.

My lab notes

Example method: Calibration of pH probe

① Use deionised water to wash the pH probe thoroughly. Shake gently to remove excess water.

② Put the pH probe into a buffer solution of known pH and record the reading.

③ Repeat for five different pH values.

④ Plot a graph of the recorded pH reading (*x*-axis) against the pH of the buffer solution (*y*-axis).

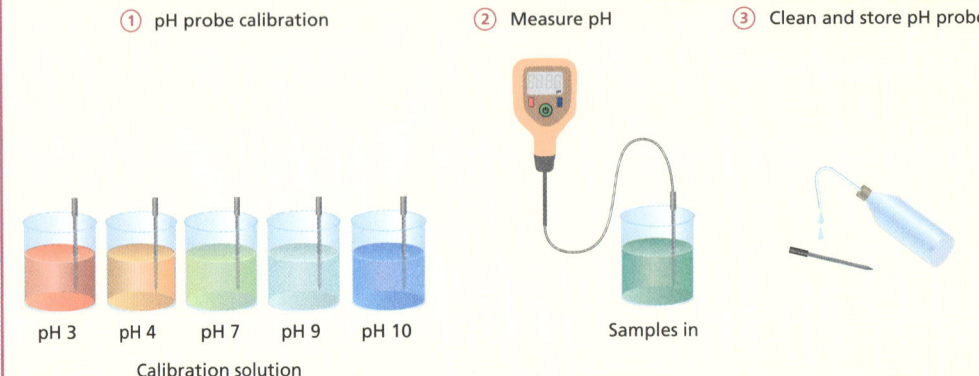

① pH probe calibration ② Measure pH ③ Clean and store pH probe

pH 3   pH 4   pH 7   pH 9   pH 10

Calibration solution

Samples in

Example method: Producing a titration curve

① Fill the burette with an acid of known concentration.

② Add $20.0\,cm^3$ of alkali to a beaker. Put the beaker on a magnetic stirring plate and add a magnetic stirrer bar.

③ Clamp a pH probe into the beaker, making sure that it will not interact with the magnetic stirrer.

④ From the burette, add $1.0\,cm^3$ of solution at a time. After each addition, measure and record the pH value.

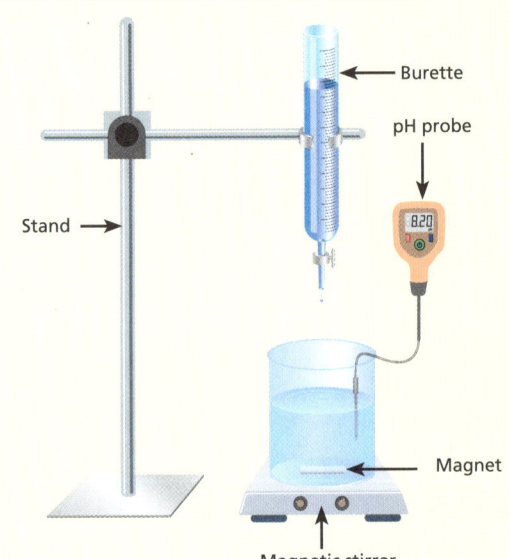

Burette

pH probe

Stand

Magnet

Magnetic stirrer

Analysis

Use the calibration curve to change each recorded pH value into the accurate pH value.

Plot a graph of the corrected pH values (*y*-axis) against the volume of acid added (*x*-axis) and draw a line of best fit.

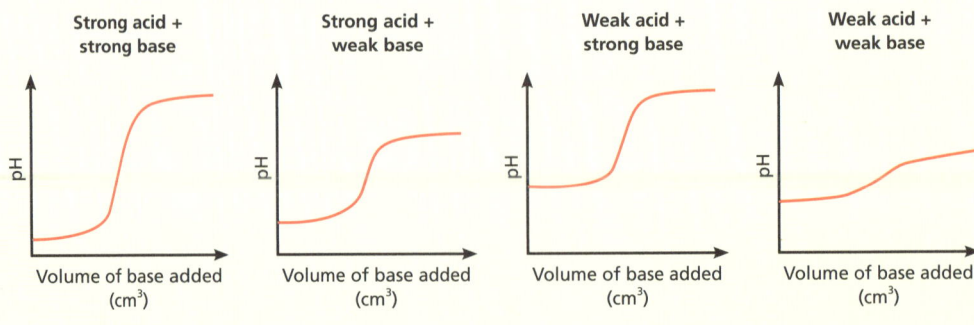

Strong acid + strong base

Strong acid + weak base

Weak acid + strong base

Weak acid + weak base

pH

Volume of base added (cm³)

Aim

To prepare a pure organic solid and test its purity, and to prepare a pure organic liquid.

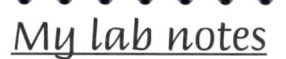

My lab notes

Example method: Preparing a sample of pure, dry aspirin

① Add 2.0 g of 2-hydroxybenzoic acid, 4 cm³ of ethanoic anhydride and 5 drops of 85% phosphoric(V) acid to a flask. Swirl to mix.

② Reflux for about 5 minutes.

③ Carefully add 2 cm³ of water.

④ Once the reaction has stopped, pour the mixture into 40 cm³ of cold water and hold in an ice bath to crystallise the product.

⑤ Purify by adding the crude product to 15 cm³ of ethanol in a boiling tube and heat in a water bath at about 75°C.

⑥ Once the aspirin has dissolved, filter the hot solution and collect the filtrate.

⑦ Allow the filtrate to cool in an ice bath.

⑧ Filter under vacuum filtration and collect the residue.

⑨ Allow the solid to dry in a drying oven and record the mass of the dry, purified solid.

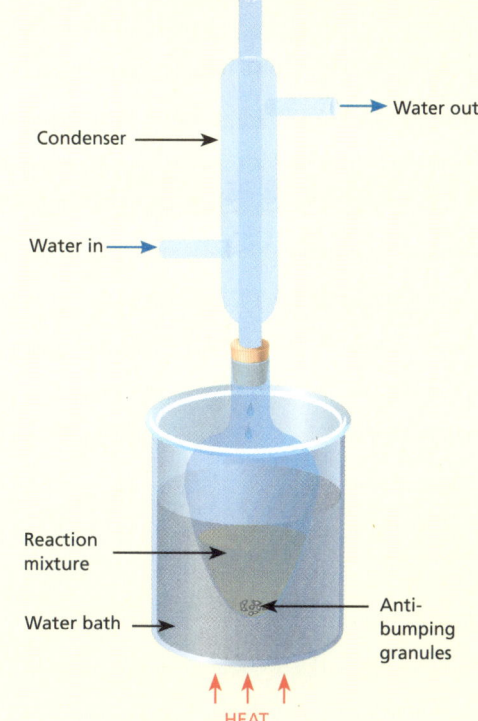

Example method: Preparation of an ester

① Add 10 cm³ ethanol, 12 cm³ ethanoic acid and 15 drops of concentrated sulfuric acid to a flask. Connect to a condenser.

② Gently heat the mixture and stop heating when no more organic liquid is vaporising.

③ Collect the distillate and add sodium carbonate and calcium chloride to the solution to remove any impurities.

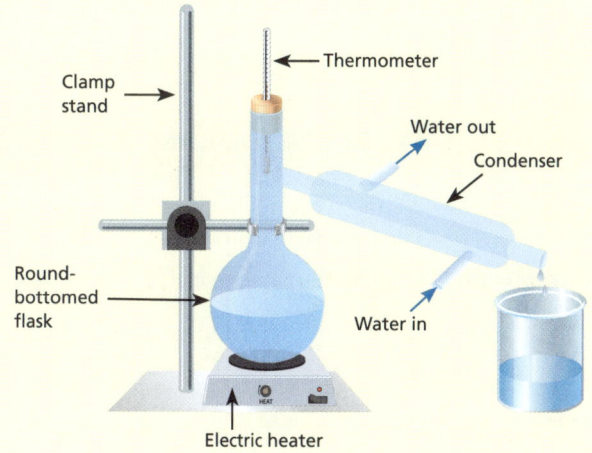

Analysis

The mass of the organic product can be measured and the percentage yield can be calculated.

The purity of the product can be determined by measuring the melting point or the boiling point. A specific melting point is a pure product. The larger the range of melting or boiling points, the more impurities there are.

Required practical 11: Identifying transition metals

Aim

To carry out simple test-tube reactions to identify transition metal ions in aqueous solution.

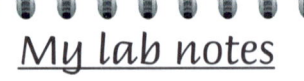

My lab notes

Example method

Test 1

① Put 10 drops of the unknown solution of transition metal into a test tube.

② Add sodium hydroxide solution dropwise until excess, shake and observe.

③ Put the test tube in a hot water bath and observe after 10 minutes.

Test 2

Put 10 drops of the unknown solution of transition metal and 10 drops of sodium carbonate solution into a test tube.

Test 3

Put 10 drops of the unknown solution of transition metal and 10 drops of silver nitrate solution into a test tube. Observe after 10 minutes.

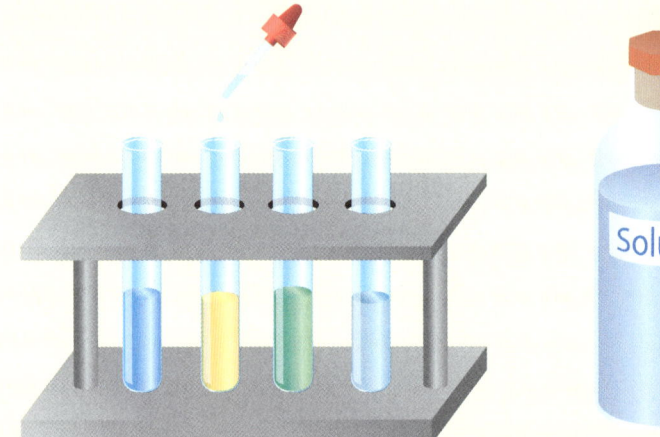

Analysis

Metal	Aqueous ion	Action of NaOH	Action of excess NaOH(aq)	Action of Na_2CO_3(aq)
Iron(II)	$[Fe(H_2O)6]^{2+}$(aq) **Green** solution	$Fe(H_2O)_4(OH)_2$(s) **Green** precipitate turns **brown** on standing in air	No further change	$FeCO_3$(s) **Green** precipitate
Copper(II)	$[Cu(H_2O)_6]^{2+}$(aq) **Blue** solution	$Cu(H_2O)_4(OH)_2$(s) **Blue** precipitate	No further change	$CuCO_3$(s) **Blue-green** precipitate
Iron(III)	$[Fe(H_2O)_6]^{3+}$(aq) **Purple** solution (may look **yellow-brown** owing to some $[Fe(H_2O)_5(OH)]^{2+}$(aq))	$Fe(H_2O)_3(OH)_3$(s) **Brown** precipitate (may look **orange-brown**)	No further change	$Fe(H_2O)_3(OH)_3$(s) **Brown** precipitate (may look **orange-brown**) and CO_2 gas evolved
Aluminium(III)	$[Al(H_2O)_6]^{3+}$(aq) Colourless solution	$Al(H_2O)_3(OH)_3$(s) White precipitate	$[Al(OH)_4]^-$(aq) Colourless solution	$Al(H_2O)_3(OH)_3$(s) White precipitate and CO_2 gas evolved

Required practical 12: Thin-layer chromatography

Aim

To separate species by thin-layer chromatography (TLC).

My lab notes

Example method: To analyse medicinal preparations (tablets)

① Prepare the sample – use a pestle and mortar to crush the tablet and dissolve in ethanol.

② Draw a pencil line approximately 1 cm above the bottom of a TLC plate. Mark equally spaced crosses along the line and label them to identify the preparation.

③ Use a capillary tube to add a sample of each preparation to separate crosses.

④ In a developing chamber, add a small amount of solvent and lower the TLC plate so that the bottom edge is just in contact with the solvent.

⑤ Place the lid on the chamber and allow the chromatogram to develop.

⑥ Remove the plate when the solvent is about 1 cm from the top of the plate and mark the solvent front with a pencil.

⑦ Allow the plate to dry. Use ultraviolet light to visualise the spots and draw around them with pencil.

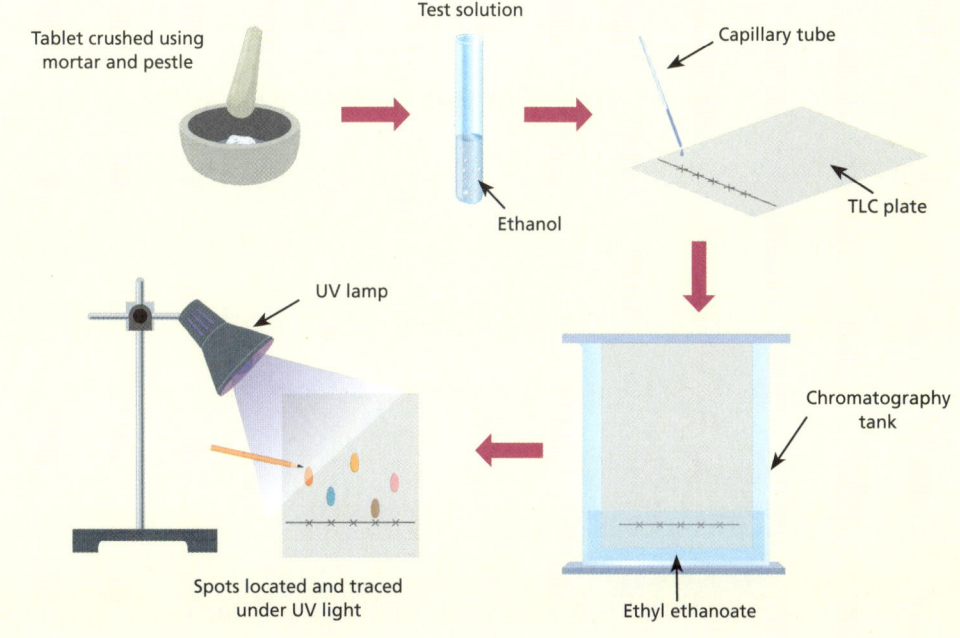

Analysis

Calculate the R_f value of each spot.

THIS PAGE HAS DELIBERATELY BEEN LEFT BLANK

Collins

Name: ....

A-level
Chemistry
Practice paper for AQA

Paper 1

Inorganic and Physical Chemistry

Time allowed: 2 hours

Materials

For this paper you must have:
- the Periodic Table/Data Booklet, see pages 218-220
- a ruler with millimetre measurements
- a scientific calculator, which you are expected to use where appropriate

Use black ink or black ball-point pen.
- Answer **all** questions.
- You must answer the questions in the spaces provided. Do not write outside the box around each page or on blank pages.
- All working must be shown.
- Do all rough work in this book. Cross through any work you do not want to be marked.

The marks for questions are shown in brackets.
- The maximum mark for this paper is 105.

Answer **all** questions in the spaces provided.

0 1 This question is about the properties of some d-block elements.

0 1 . 1 Give the full electron configuration of a zinc atom.

[1 mark]

....

0 1 . 2 Explain why zinc is **not** a transition metal.

[2 marks]

....

....

0 1 . 3 In the space below, draw a diagram of the $[Cu(H_2O)_2(NH_3)_4]^{2+}$(aq). Give the shape of the complex ion.

[2 marks]

0 1 . 4 When concentrated HCl is added to a solution containing $[Co(H_2O)_6]^{3+}$ a ligand exchange reaction occurs.

Write a balanced equation for this reaction.

[1 mark]

....

0 2 This question concerns ionisation energies.

0 2 . 1 Explain the meaning of the term 'first ionisation' of an atom.

[1 mark]

....

....

0 2 . 2 State and explain the general trend in first ionisation energies across Period 3 of the Periodic Table, Na–Ar.

[2 marks]

Trend

....

Explanation

....

....

0 2 . 3 Complete **Figure 1** to show the general trend of the successive ionisation energies for Na.

[2 marks]

Figure 1

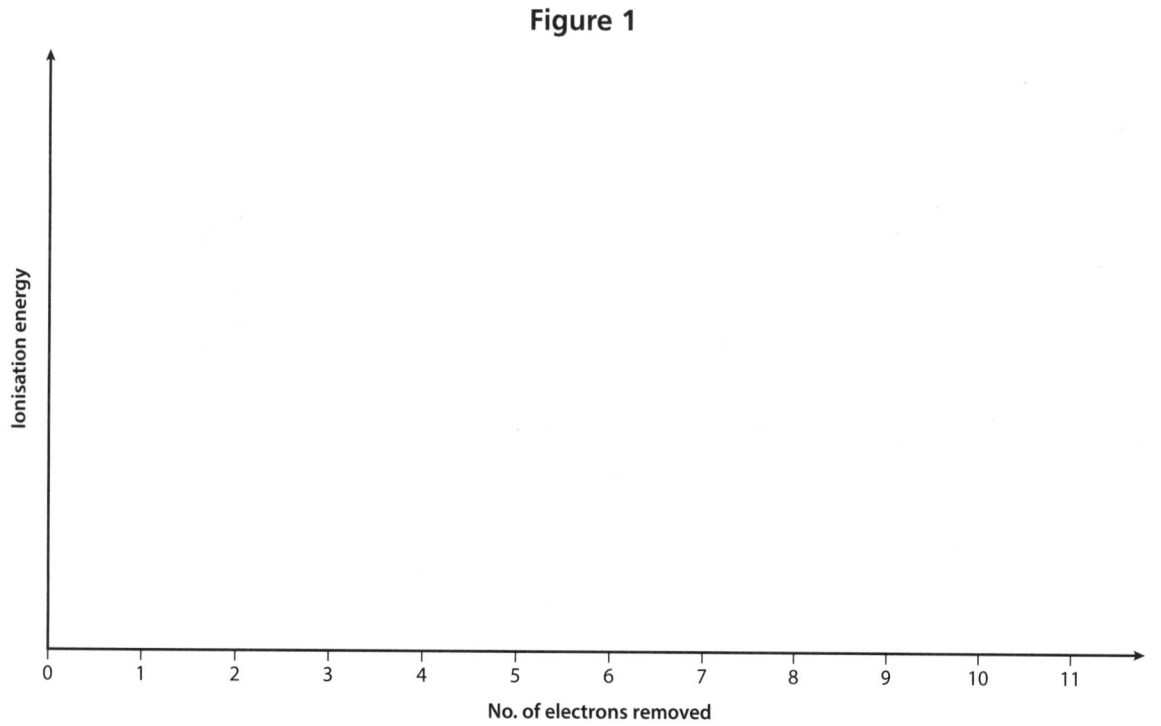

0 2 · 4 Explain how the isotopic abundance of an element is measured using a time of flight (TOF) mass spectrometer.

You should include a description of each of the main stages in the process.

[4 marks]

0 2 · 5 A sample of rubidium contains three isotopes, ^{85}Rb and ^{87}Rb.

Calculate the percentage abundance of each isotope.

Give your answer to 1 decimal place.

[1 mark]

0 3   Magnesium sulfate can also occur as a hydrated compound, $MgSO_4 \cdot xH_2O$

A student uses this method to determine the value of X in hydrated magnesium sulfate.

1. Add 10.0 g of hydrated magnesium sulfate and 100.00 cm³ of distilled water into a beaker.
2. Stir to ensure the solid has fully dissolved.
3. Add 100.00 cm³ of 1.0 moldm⁻³ barium chloride solution and stir.
4. Filter the mixture.
5. Collect the residue and dry.
6. Record the mass of the dry residue.

The mass of the precipitate collected was 9.47 g.

0 3 · 1 Write a balanced equation, including state symbols, for the reaction between the solutions of magnesium sulfate and barium chloride.

[1 mark]

0 3 · 2 Determine which substance was in excess.

[3 marks]

0 3 · 3 Determine the number of moles of hydrated magnesium sulfate in the 10.0 g sample and hence the value of x in the formula.

[3 marks]

0 3 . 4 Another student wanted to determine the water of crystalisation in hydrated magnesium sulfate.

Suggest an alternative method for the student to use that involves heating the hydrated magnesium sulfate.

[4 marks]

0 4   The Contact Process is important for the industrial production of sulfuric acid.

The Contact Process involves an equilibrium between sulfur dioxide and sulfur trioxide.

$$2SO_2(g) + O_2(g) \rightleftharpoons 2SO_3(g) \qquad \Delta H^\theta = -196 \text{ kJ mol}^{-1}$$

An equilibrium mixture of the gases was found to contain the concentrations shown in **Table 1**.

Table 1

SO_2	3.4×10^{-9} mol dm^{-3}
O_2	2.0×10^{-8} mol dm^{-3}
SO_3	0.97 mol dm^{-3}

0 4 . 1 Calculate the value for the equilibrium constant K_c for oxidation of sulfur dioxide to form sulfur trioxide.

State the units.

[4 marks]

0 4 · 2 Which of these three statements about this equilibrium is true? Tick [✓] **one** box.

[1 mark]

A The forward reaction is favoured by a decrease in pressure. ☐

B K_c and K_p will have the same value for this equilibrium. ☐

C The forward reaction is favoured by a decrease in temperature. ☐

0 4 · 3 Which of these three statements about the rates of the forward and reverse reactions is true? Tick [✓] **one** box.

[1 mark]

A The rate of the forward reaction is unaffected by temperature change. ☐

B At equilibrium, the rate of the reverse reaction is lower than the rate of the forward reaction. ☐

C The rate of the forward and the reverse reactions will be equally affected by a change in pressure. ☐

0 4 · 4 Vanadium(V) oxide acts as a heterogeneous catalyst in the Contact Process.

Explain the effect on rate and yield of the use of a catalyst in the Contact Process.

[2 marks]

....

....

0 4 · 5 Any waste gases from the Contact Process can cause air pollution. Describe one environmental problem caused by sulfur dioxide gas.

Give an equation to show the production of the environmental problem from the sulfur dioxide.

[2 marks]

....

....

0 5   This question is about Group 7 elements, the halogens.

0 5 · 1 Define the term *electronegativity*.

[1 mark]

0 5 · 2 Explain the trend in electronegativities shown by the halogens.

[2 marks]

0 5 · 3 i) An equation for the reaction that takes place when chlorine gas is bubbled through aqueous potassium iodide. Write an ionic equation for this reaction.

[1 mark]

ii) Determine which species is the oxidising agent.

[1 mark]

Table 2 shows the boiling points of four of the halogens.

Table 2

Element	Boiling point (K)
Fluorine	53.5
Chlorine	171.7
Bromine	265.9
Iodine	386.8

0 5 . 4 Give an equation to show the boiling of liquid bromine.

[1 mark]

0 5 . 5 Explain why bromine has a higher boiling point than fluorine.

[2 marks]

0 5 . 6 Iodine changes directly from a solid to a gas at 386.8 K. Give the name of this state change.

[1 mark]

0 6   This question is about the oxides of Period 3 elements.

0 6 · 1 Describe how the acid-base characteristics of the Period 3 oxides change as you go across the period from Na → Cl.

[2 marks]

_____

_____

_____

_____

0 6 · 2 Give the equation for the reaction between sodium oxide and water.

[1 mark]

_____

0 6 · 3 Give the equation for the reaction of phosphorous(V) oxide and water.

[1 mark]

_____

0 6 · 4 **Table 3** below shows the melting point of three Period 3 oxides.

Table 3

Period 3 oxide	Melting point / K
Na_2O	1193
SO_2	198
SiO_2	1883

Explain why the melting points of these substances are different.

You should refer to the structure of and bonding in each substance.

[6 marks]

_____

_____

_____

_____

_____

_____

_____

0 6 · 5 Silicon dioxide reacts with calcium oxide as wet concrete hardens. Give an equation for the reaction of silicon dioxide with calcium oxide.

[1 mark]

_____

0 7   This question is about titration.

A student dissolved an unknown mass made of two sodium hydroxide pellets in water to make a 250 cm³ solution. A 25.0 cm³ sample of this sodium hydroxide solution is placed in a conical flask and is titrated with 0.100 mol dm⁻³ hydrochloric acid. The two concordant titre values were **28.80 cm³** and **28.70 cm³**.

0 7 · 1 Give an equation, including state symbols, for the reaction between hydrochloric acid and sodium hydroxide.

[1 mark]

0 7 · 2 Hydrochloric acid is a strong acid and sodium is a strong base. Define the term 'strong'.

[1 mark]

0 7 · 3 Determine the number of moles of hydrochloric acid needed to neutralise 25.00 cm³ of the sodium hydroxide solution.

[2 marks]

0 7 . 4 Calculate the concentration of the sodium hydroxide solution, in mol dm⁻³.

[3 marks]

0 7 . 5 Use the results to calculate the mean mass of a sodium hydroxide pellet.

[3 marks]

0 8 This question is about thermodynamics.

0 8 · 1 Consider the reaction shown:

$C_2H_6(g) \rightarrow C_2H_4(g) + H_2(g)$

Table 4 shows some thermodynamic data.

Table 4

Standard enthalpy of combustion	Value / kJ mol^{-1}
$C_2H_6(g)\ \Delta H_c^\theta$	–1560
$H_2(g)\ \Delta H_c^\theta$	–286
$C_2H_4(g)\ \Delta H_c^\theta$	–1411

Use the data from the table to calculate the enthalpy change of this reaction. You may wish to draw a suitable cycle.

[2 marks]

....

0 8 · 2 The standard entropy for some substances are given.

$C_2H_6(g)\ S^\theta = 230\ J\ K^{-1}\ mol^{-1}$

$H_2(g)\ \ \ S^\theta = 131\ J\ K^{-1}\ mol^{-1}$

$C_2H_4(g)\ S^\theta = 220\ J\ K^{-1}\ mol^{-1}$

Calculate the entropy change, ΔS, for the reaction: $C_2H_6(g) \rightarrow C_2H_4(g) + H_2(g)$

[1 mark]

....

0 8 · 3 Explain what a positive entropy change means.

[2 marks]

....

....

....

....

0 8 · 4 Use the answers you have calculated in parts **8.1** and **8.3** to calculate the Gibbs free energy change, ΔG^{θ}, for the reaction: $C_2H_6(g) \rightarrow C_2H_4(g) + H_2(g)$

[2 marks]

....

0 8 · 5 Suggest how temperature affects the feasibility of the reaction.

[1 mark]

....

....

....

0 8 · 6 The value for the enthalpy change of the reaction

$C_2H_6(g) \rightarrow C_2H_4(g) + H_2(g)$, suggests that

Tick [✓] **one** box.

[1 mark]

A   the bonds broken in this reaction are stronger than the bonds made.

B   the bonds made in this reaction are stronger than the bonds broken.

C   ethene is more stable than ethane.

0 9 This question is about preparation of a buffer solution.

0 9 · 1 State the meaning of the term 'buffer solution'

[2 marks]

0 9 · 2 A buffer solution can be made from the partial neutralisation of a weak acid with a strong base. A student made an acidic buffer solution by reacting 100.00 cm³ 0.1 moldm⁻³ propanoic acid solution with 100 cm³ 0.05 moldm⁻³ sodium hydroxide solution.

Give an equation for equilibrium in the buffer solution.

[1 mark]

0 9 · 3 Explain how this buffer solution resists changes in pH. Use equations in your answer.

[2 marks]

0 9 · 4 Calculate the pH of a 0.100 mol dm^{-3} solution of propanoic acid at 25°C.

Give your answer to 2 decimal places.

[2 marks]

0 9 · 5 Determine the pH of this acid buffer solution at 25°C made from propanoic acid and sodium hydroxide solution.

The K_a of propanoic acid is 1.3×10^{-5} mol dm^{-3}.

[3 marks]

1 0 This question is about fuel cells.

1 0 . 1 In a methanol-oxygen fuel cell, a REDOX reaction occurs with a EMF of +1.20V.

Hydrogen ions react with oxygen to form water.

Methanol reacts with water to form carbon dioxide and hydrogen ions.

a. Deduce the half equation for the reaction at each electrode.

[2 marks]

(i) Anode

(ii) Cathode

b. Give the ionic equation for the overall reaction in the methanol-oxygen fuel cell.

[1 mark]

c. State why a fuel cell does not need to be electrically recharged.

[1 mark]

d. Explain how to keep the fuel cell EMF constant.

[2 marks]

e. Suggest advantages of using methanol as a fuel instead of hydrogen as a fuel for cars.

[2 marks]

f. Suggest why the EMF values of the acidic and alkaline hydrogen–oxygen fuel cells are the same.

[1 mark]

g. Suggest why ethanol fuel cells can be considered to be using a carbon-neutral fuel.

[1 mark]

1 0 . 2 A student wanted to investigate the products of electrolysis of a dilute solution of sulfuric acid.

Draw a labelled diagram of the equipment that could be used in this investigation.

[4 marks]

1 0 . 3 Give the half equations for the reactions that occur at each electrode.

[2 marks]

Anode ....

Cathode ....

1 0 . 4 Explain the changes in concentration of sulfuric acid as the electrolysis reaction occurs.

[3 marks]

....

....

....

....

END OF QUESTIONS

Collins

Name:

A-level
Chemistry
Practice paper for AQA

Paper 2

Organic and Physical Chemistry

Time allowed: 2 hours

Materials

For this paper you must have:
- the Periodic Table/Data Booklet,
  see pages 218-220
- a ruler with millimetre measurements
- a scientific calculator, which you are
  expected to use where appropriate

Use black ink or black ball-point pen.
- Answer **all** questions.
- You must answer the questions in the spaces provided. Do not write outside the box around each page or on blank pages.
- All working must be shown.
- Do all rough work in this book. Cross through any work you do not want to be marked.

Information
- The maximum mark for this paper is 105.

Answer **all** the questions in the spaces provided.

0 1 Under alkali conditions, hydrolysis of 1-bromobutane forms an alcohol organic product.

Table 1 shows how the values of the relative initial rate vary with different concentrations of each reagent at 298K.

Table 1

Experiment	$[C_4H_9Br]$	$[OH^-]$	Initial rate (mol dm^{-3} s^{-1})
1	0.050	0.10	4.0×10^{-4}
2	0.15	0.10	1.2×10^{-3}
3	0.10	0.20	1.6×10^{-3}

0 1 . 1 Deduce the order of reaction with respect to 1-bromobutane. Give a reason for your answer.

[2 marks]

Order

....

Reason

....

....

....

0 1 . 2 Deduce the order of reaction with respect to the hydroxide ion. Give a reason for your answer.

[2 marks]

Order

....

Reason

....

....

....

0 1 · 3 Deduce the rate equation for the hydrolysis of 1-bromobutane.

[1 mark]

....

0 1 · 4 Calculate a value for the rate constant at this temperature and give its units.

[2 marks]

....

0 1 · 5 Suggest how the rate constant would change if the temperature was increased to 333K. Explain your answer.

[3 marks]

....

....

....

....

0 2   This question is about halogenated organic compounds.

0 2 · 1   Under UV light, ethane and chlorine react to form a number of chlorine-containing products, including C_2H_5Cl.

Describe how C_2H_5Cl can be formed.

[2 marks]

Give balanced equations in your answer and name the mechanism.

[2 marks]

0 2 · 2   Explain why ultraviolet light is needed for this reaction.

[1 mark]

0 2 · 3   Chlorine radicals are accelerating the ozone depletion in Earth's atmosphere. The overall equation for this reaction is:

$$2O_3(g) \rightarrow 3O_2(g)$$

Explain why chlorine radicals are not included in the overall equation for ozone depletion.

[1 mark]

0 2 · 4   Give **two** equations to show how chlorine radicals catalyse the decomposition of ozone.

[2 marks]

0 2 . 5 Explain why a single radical can cause the decomposition of many molecules of ozone.

[2 marks]

_____

_____

0 2 . 6 Halogenated organic substances are used as refrigerants. A manufacture of freezers changed their refrigerant gas from halon (CF_3CHCl_2), to norfluorane (CF_3CH_2F).

Sometimes the refrigerate gases can escape into the atmosphere.

Explain how the environmental impact of norfluorane is lower than halon.

[4 marks]

_____

_____

_____

_____

0 3 This question is about amino acids. Glycine is an amino acid with the molecular formula $C_2H_5NO_2$.

0 3 · 1 Draw the displayed formula of glycine.

[3 marks]

0 3 · 2 Draw the structure of the species formed when glycine reacts with an excess of bromomethane.

[1 mark]

0 3 · 3 Alanine is an amino acid which can form optical isomers. Draw three-dimensional representations of the two enantiomers of alanine, showing how the two structures are related to each other.

Use page 220 of the Data Booklet.

[2 marks]

0 3 · 4 State the meaning of the term 'zwitterions'.

[2 marks]

0 3 · 5 A solution of alanine can act as a buffer solution. Give equations to show how alanine can maintain the pH on additions of:

(i) small amounts of HCl

(ii) small amounts of NaOH

(i) HCl

[1 mark]

(ii) NaOH

[1 mark]

0 3 · 6 Alanine can react with glycine to make two different dipeptides. Use skeletal formulae to show the two different dipeptides.

[2 marks]

0 3 · 7 Name the technique which is used to identify amino acids in a mixture of amino acids.

[1 mark]

0 4   **Figure 1** shows the skeletal formula of an organic compound.

Figure 1

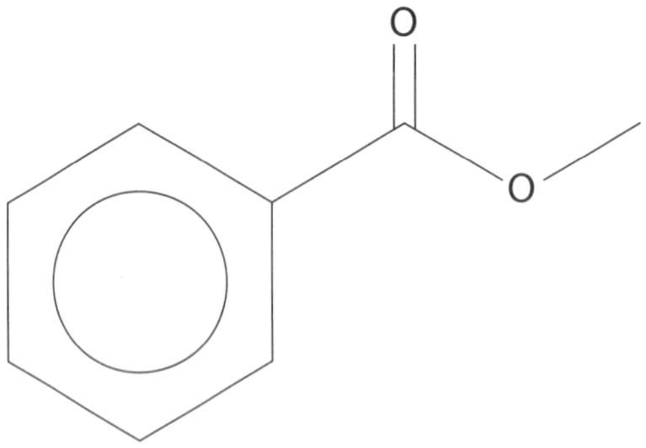

0 4 · 1 Name the organic substance shown in **Figure 1**.

[1 mark]

0 4 · 2 A student prepared a dry pure sample of 9.98 g of the ester shown in **Figure 1**. The student used 10.50 g of an organic acid and reacted it with an excess of an alcohol. Calculate the percentage yield of the organic product.

The relative formula mass for the organic acid is 122.

[3 marks]

0 4 · 3 The organic substance in **Figure 1** can undergo a nitration reaction.

Explain how the electrophile is produced.

Give an equation for the production of reactive species which attacks the benzene ring.

Name the reactive species which attacks the benzene ring.

[4 marks]

0 4 · 4 Complete the mechanism to show the nitration of methyl benzoate by adding two curly arrows and any relevant charges.

[4 marks]

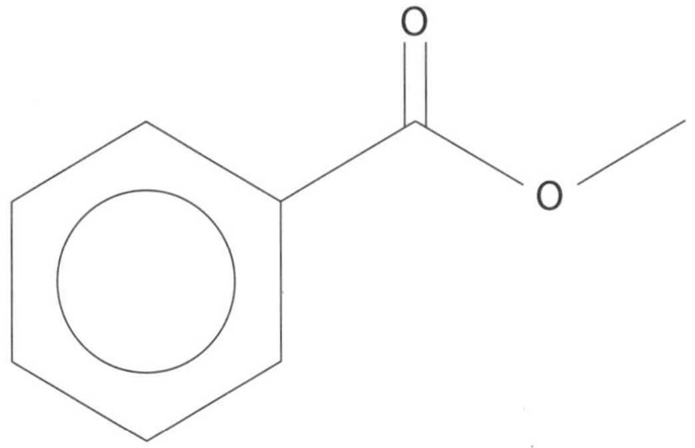

The IR spectrum of an organic substance with empirical formula C_4H_8O is shown in **Figure 2**.

Figure 2

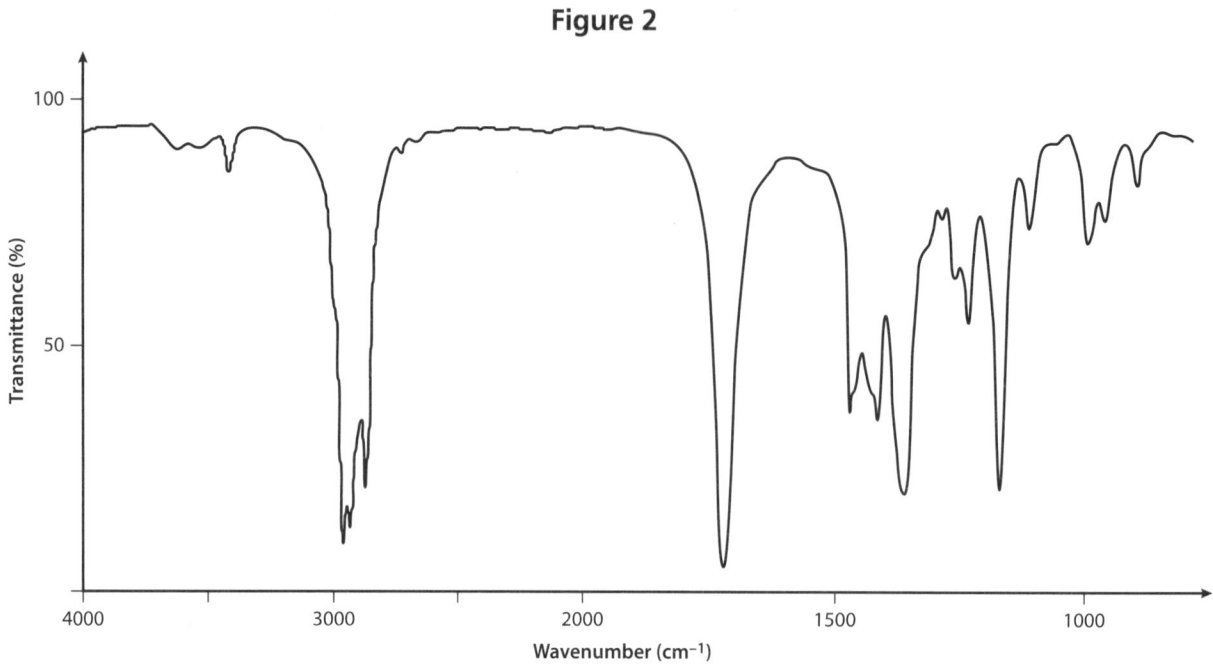

0 5 · 1 Explain how infra red spectroscopy can be used to identify bonds present in an organic substance.

[2 marks]

0 5 · 2 Use Table A on the Data Booklet to suggest the absorption range for the carbonyl functional group.

[1 mark]

0 5 · 3 Describe how the student could use a chemical test to determine if the organic substance was an aldehyde or ketone.

[3 marks]

....

....

....

0 5 · 4 Further tests showed that the organic compound was a ketone. Draw the skeletal formula and give the systematic name for this organic compound.

[2 marks]

0 6 This question is about the chlorination of alkenes.

0 6 · 1 Give an equation, including state symbols, for the reaction of ethene with chlorine.

[2 marks]

0 6 · 2 Complete the mechanism to show the chlorination of ethene by adding three curly arrows and any relevant charges.

[4 marks]

Figure 3

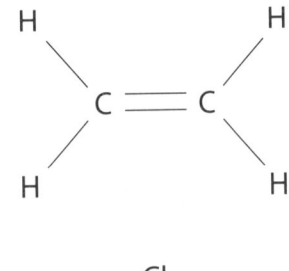

0 6 · 3 Name the organic product of this chlorination reaction.

[1 mark]

0 6 · 4 Two stereoisomers are formed by the chlorination of propene. Give the structures of these two isomers and name the type of stereoisomerism shown.

Explain how this isomerism is possible.

[4 marks]

0 6 · 5 Describe how ^{1}H NMR spectroscopy could distinguish between 1,2-dichloropropane and 2,2-dichloropropane.

[2 marks]

0 7 This question is about synthetic polymers.

Polyethene is is made from a single monomer. Terylene is made from two different monomers, ethane-1,2-diol and benzene-1, 4-dicarboxylic acid.

0 7 . 1 Draw a repeating unit of terylene. You should include four monomers.

[2 marks]

0 7 . 2 Explain how the polymerisation process is different between ethene and terylene.

[4 marks]

0 7 · 3 Chemical storage containers can be from polyethene or Terylene. Explain why a Terylene plastic bottle would not be suitable for storing sodium hydroxide solution. Use your knowledge of structure and bonding.

[3 marks]

| 0 | 8 | | This question is about organic synthesis.

Substance Q is an organic compound that is used in the food industry as strawberry flavouring.

Figure 4

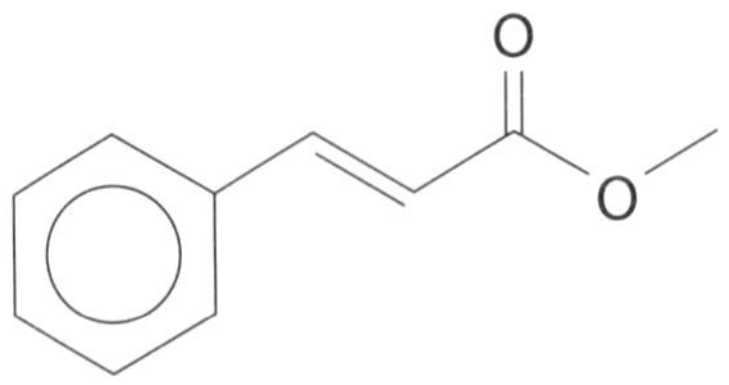

Substance Q

Substance P is a naturally occurring aldehyde that can be used to make the strawberry flavouring.

Figure 5

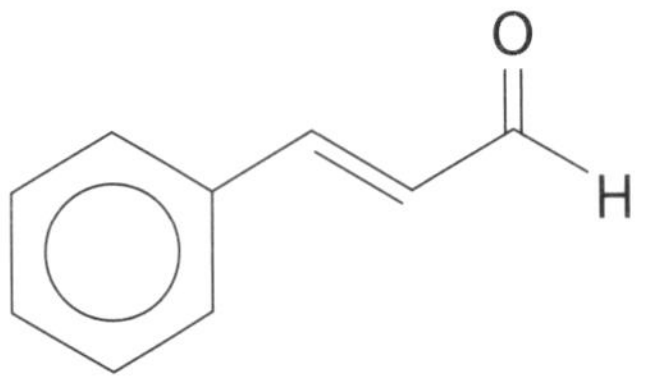

Substance P

| 0 | 8 | · | 1 | Suggest a two-stage synthesis for making substance Q using P as a starting material. Name all the reagents used and draw the structure of intermediate products.

[4 marks]

Stage 1: reagent(s) ....

Structure of intermediate

Stage 2: reagent(s) ....

0 8 · 2 Each step in the two-stage synthetic route has a yield of 60.00%. In a small test laboratory, a process chemist wanted to make and collect 100 g of pure dry organic substance Q. Calculate the mass of organic substance needed. The relative formula mass (M_r) of P is 132.

[5 marks]

0 8 · 3 An alternative synthesis of Q involves the reaction between two starting materials, A and B.

[2 marks]

Figure 6

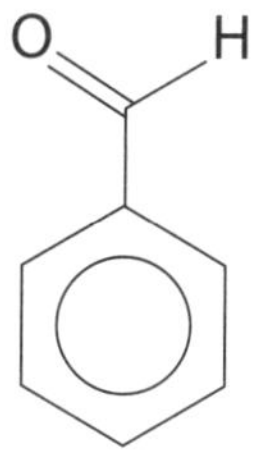

A 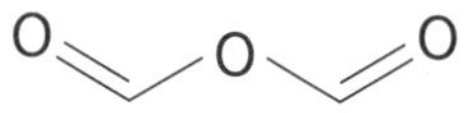   B

Name starting materials A and B, used in this synthesis.

A ....

B ....

0 8 · 4 Give **two** examples of important considerations when chemists design synthetic routes to produce organic substances. You do not need to mention cost.

[2 marks]

Seligine is a drug which is used to treat depression. Seligine inhibits enzymes which act on neurotransmitters. **Figure 7** below shows the skeletal formula of the biologically active organic substance in Selginine.

Figure 7

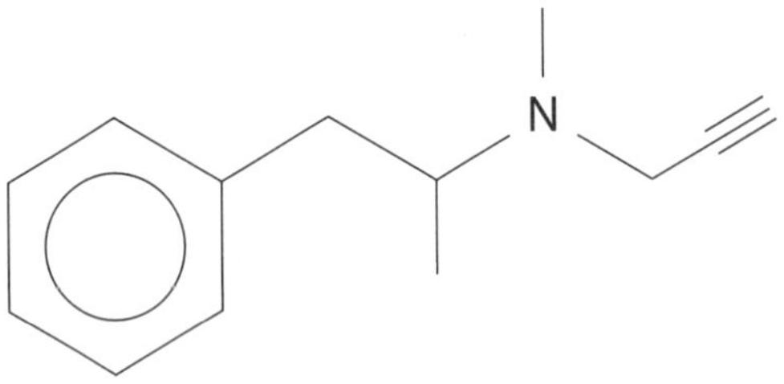

0 9 · 1 Determine if Seliginine is a primary, secondary or tertiary amine.

[1 mark]

0 9 . 2 Suggest how Seliginine inhibits the enzyme monoamine oxidase.

[2 marks]

....

....

....

....

....

0 9 . 3 Which type of bond holds together the primary structure of a protein? Tick [✓] one box.

[1 mark]

A   Covalent bond   ☐

B   Hydrogen bond   ☐

C   Ionic bond   ☐

0 9 . 4 Which type of bond maintains the secondary structure of a protein? Tick [✓] **one** box.

[1 mark]

A Disulfide bridge (S–S bond)

B Hydrogen bond

C Induced dipole-dipole forces

0 9 . 5 Describe **two** differences between a protein molecule and a molecule of DNA.

[2 marks]

1 0   An organic substance Z has a relative molecular mass of 150

1 0 . 1 A sample of Z is found to contain 37.08% carbon and 6.7% hydrogen by mass. The rest of the sample was oxygen. Calculate the empirical formula of Z.

[3 marks]

molecular formula _____

1 0 . 2 Z has both a carboxylic acid group and a phenyl group. The ^{13}C NMR spectrum for Z shows only seven peaks from seven different carbon environments.

Draw the possible isomers of Z that fit this information.

[3 marks]

THIS PAGE HAS DELIBERATELY BEEN LEFT BLANK

Collins

A-level
Chemistry
Practice paper for AQA

Paper 3

Time allowed: 2 hours

Materials

For this paper you must have:
- the Periodic Table/Data Booklet, see pages 218-220
- a ruler with millimetre measurements
- a scientific calculator, which you are expected to use where appropriate

Use black ink or black ball-point pen.
- Answer **all** questions.
- You must answer the questions in the spaces provided. Do not write outside the box around each page or on blank pages.
- All working must be shown.
- Do all rough work in this book. Cross through any work you do not want to be marked.

The marks for questions are shown in brackets.
- The maximum mark for this paper is 90.

Advice
- You are advised to spend 70 minutes on Section A and 50 minutes on Section B.

Section A

Answer **all** questions in this section.

---

0 1 This question is about preparing a standard solution.

A student makes a 0.10 mol dm^{-3} standard solution of sodium hydrogensulfate, $NaHSO_4$.

0 1 · 1 Calculate the mass of sodium hydrogensulfate the student needs to use to make a 250.00 cm^3 volume of 0.10 moldm^{-3} solution of sodium hydrogensulfate.

[2 marks]

0 1 · 2 The student uses a top-pan balance accurate to two decimal places. The student measures the weighing boat and the sodium hydrogencarbonate and the weighing boat after they have transfered the sodium hydrogensulfate to a 100 cm^3 glass beaker.

Explain how the student will accurately determine the mass of sodium hydrogensulfate used.

[2 marks]

0 1 · 3 Use your answer to 1.1. to calculate the percentage error of the measurement of the mass of sodium hydrogensulfate.

If you do not have an answer to 1.1. use the value 4.0 g. This is not the correct answer.

[3 marks]

$\boxed{0}\ \boxed{1}\cdot\boxed{4}$ The student then adds a small amount of distilled water and stirs until the solid sodium hydrogensulfate has fully dissolved. The student then transfers the solution into a 250 cm³ volumetric flask.

Describe how the student ensures that as much of the sodium hydrogen sulfate is transferred.

[2 marks]

$\boxed{0}\ \boxed{1}\cdot\boxed{5}$ The student added water into the volumetric flask to make up to 25.00 cm³ of solution.

(i) Explain how the student ensures that exactly 25.00 cm³ of solution is made.

(ii) Suggest one improvement to this procedure.

[4 marks]

0 2 This question is about electrochemical cells.

Figure 1 shows how a student combined a zinc half-cell and a copper half-cell to measure the EMF.

Figure 1

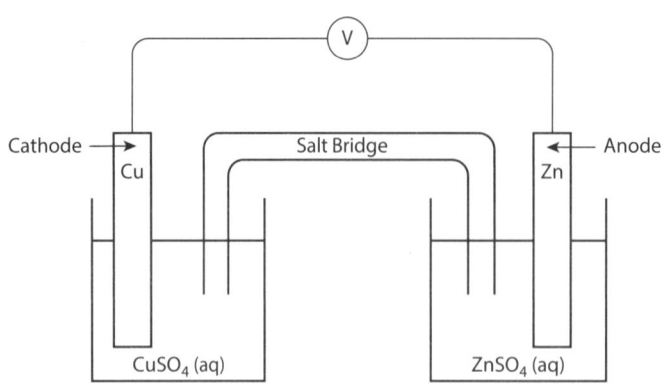

0 2 · 1 Give equations for the equilibria in each half-cell.

[2 marks]

Zinc half-cell

....

Copper half-cell

....

0 2 · 2 Explain why the zinc is the negative electrode (anode) in this cell.

[2 marks]

....

....

....

0 2 · 3 The standard electrode potentials for copper and zinc are shown below.

$$E^\theta \ Cu = +0.34 \ V$$

$$E^\theta \ Zn = -0.76 \ V$$

Calculate the EMF of this cell under standard conditions.

Give the standard conditions.

[2 marks]

....

| 0 | 2 |·| 4 | Suggest the effect on the cell EMF if distilled water was added to the zinc half-cell.

[1 mark]

....

| 0 | 2 |·| 5 | State the purpose of a salt bridge.

[1 mark]

....

| 0 | 2 |·| 6 | Explain why electrochemical cells can be described as an example of a 'redox' reaction.

Use this electrochemical cell as an example in your answer

[3 marks]

....

....

....

....

....

0 3   A student does an experiment to determine a value for the enthalpy of reaction for a displacement reaction between copper(II) sulfate and magnesium powder.

The student uses this method.

- Measure 25.0 cm³ of 0.05 moldm⁻³ copper(II) sulfate.
- Pour the copper(II) sulfate into a beaker.
- Record the temperature of the copper(II) sulfate in the beaker.
- Add 0.50 g of powdered magnesium to the copper(II) sulfate in the beaker.
- Stir the solution and record the highest temperature reached.

Table 1 shows the student's results.

Table 1

Initial temperature of solution (°C)	21.0
Maximum temperature of solution (°C)	26.5

0 3 · 1 Name a piece of equipment which can accurately measure the volume of copper(II) sulfate solution.

[1 mark]

....

0 3 · 2 Give an ionic equation for the reaction.

[1 mark]

....

0 3 · 3 Copper(II) sulfate is the limiting reactant. Calculate the amount, in moles, of magnesium used in this practical.

[3 marks]

....

0 3 . 4 Calculate the enthalpy change, in kJmol⁻¹, for the copper(II) sulfate in this experiment.

Use the data in **Table 1**.

Assume that the specific heat capacity of the solution, $c = 4.18$ J K⁻¹ g⁻¹

Assume that the density of the solution = 1.00 g cm⁻³

[2 marks]

0 3 . 5 Suggest an improvement to the method to reduce the main source of error.

[2 marks]

0 3 . 6 Suggest the effect on enthalpy change if 1.0 g of magnesium powder was used with 25.00 cm³ 0.5 oldm⁻³ copper(II) sulfate solution.

[2 marks]

Effect

Explanation

0 4 Ethanol can be prepared by the oxidation of ethanol by acidified potassium dichromate.

The reaction mixture is gently heated under reflux and the product is collected by distillation.

0 4 · 1 Give the equation for the oxidation of ethanol to produce ethanal. Use [O] for the oxidising agent.

[1 mark]

0 4 · 2 Suggest one improvement to the method to prevent the reaction mixture boiling over into a collection tube.

[2 marks]

0 4 · 3 The distillate is collected in a conical flask which is standing in an ice bath made from a beaker of iced water. Explain the purpose of the ice bath.

[2 marks]

0 4 · 4 Describe **one** chemical test that the student could carry out to show that the distillate contained ethanal. Include the expected result.

[2 marks]

0 4 · 5 One impurity that may be present in the distillate is ethanoic acid. Suggest how ethanoic acid could have been produced and how ethanoic acid can be removed.

[3 marks]

0 4 · 6 A student oxidised 4.60 g of ethanol and collected 3.80 g of ethanol. Calculate the percentage yield.

[4 marks]

| 0 5 | This question is about the techniques used in titration.

A student titrates a solution of ethanoic acid with a standard solution of sodium hydroxide. Phenolphthalein was used as an acid-base indicator and the rough titration showed that the concentration of sodium hydroxide was about 0.01 moldm⁻³. The accurate titrations were completed using a pH probe and a pH curve was plotted for each titration. The equivalence point was determined from the graph and used in calculations to determine the accurate concentration of the sodium hydroxide solution.

| 0 5 |·| 1 | Suggest the equipment used to measure the volume of the sodium hydroxide solution and the volume of the ethanoic acid solution.

[2 marks]

Burette ....

Pipette ....

| 0 5 |·| 2 | Explain why phenolphthalein is a suitable indicator for the titration between sodium hydroxide and ethanoic acid.

[5 marks]

0 5 . 3 Describe how to use a pH curve to determine the equivalence point.

[2 marks]

0 5 . 4 Explain why the pH of the titration mixture remains approximately constant when 10.00 cm³ to 15.00 cm³ of NaOH is added.

[2 marks]

Section B

Answer **all** questions in this section.

Only **one** answer per question is allowed.

For each answer completely fill in the circle alongside the appropriate answer.

CORRECT METHOD ● WRONG METHODS ⊗ ⊙ ⊜ ✓

If you want to change your answer, you must cross out your original answer as shown. ⊠

If you wish to return to an answer previously crossed out, ring the answer you now wish to select as shown. ⊗

You may do your working in the blank space around each question but this will not be marked. Do **not** use additional sheets for this working.

`0 6` What is the role of chlorine when it reacts with benzene?

[1 mark]

A an electrophile ◯

B a nucleophile ◯

C a reducing agent ◯

D a free radical ◯

`0 7` Which of these is **not** a reason why potassium is a more reactive element than sodium?

[1 mark]

A Potassium's outer shell electron is further from the nucleus ◯

B Potassium has a more positively charged nucleus ◯

C Potassium has a lower first ionisation energy than sodium ◯

D In a potassium atom, the outer shell electron experiences more shielding ◯

0 8 5.00 kg of calcium carbonate produces 2.50 kg of calcium oxide by thermal decomposition. What volume of carbon dioxide would be produced (measured at 298 K and 100 kPa)?

[1 mark]

A   1200.0 dm^3

B   600.0 dm^3

C   714.2 dm^3

D   1071.4 dm^3

0 9 Which of the following will be different between enantiomers of an organic substance?

[1 mark]

A   melting point

B   rotation of plane-polarised light

C   reactivity

D   solubility

1 0 Which electron arrangement represents halogen?

[1 mark]

A $1s^2 2s^2 2p^6 3s^2 3p^6 4s^1$

B $1s^2 2s^2 2p^6 3s^2 3p^6 4s^2 3d^7$

C $1s^2 2s^2 2p^6 3s^2 3p^6 4s^2 3d^{10} 4p^4$

D $1s^2 2s^2 2p^6 3s^2 3p^6 4s^2 3d^{10} 4p^5$

1 1 Which of these has the compounds arranged in order of increasing boiling point?

[1 mark]

A $CH_3CH_2CH_2CH_3$, $CH_3CH(OH)CH_3$, CH_3CH_2COOH, CH_3COCH_3

B CH_3COCH_3, $CH_3CH(OH)CH_3$, CH_3CH_2COOH, $CH_3CH_2CH_2CH_3$

C $CH_3CH_2CH_2CH_3$, CH_3COCH_3, $CH_3CH(OH)CH_3$, CH_3CH_2COOH

D CH_3CH_2COOH, $CH_3CH_2CH_2CH_3$, CH_3COCH_3, $CH_3CH(OH)CH_3$

1 2 Which of these oxides would have the lowest pH?

[1 mark]

A Na_2O

B Al_2O_3

C SO_2

D SiO_2

1 3 A chemical reaction can be described by the rate equation $r = k[A][B]^2$, what are the units of k?

[1 mark]

A $mol^2\ dm^{-6}\ s^{-1}$

B $mol^{-2}\ dm^6\ s^{-1}$

C $mol^3\ dm^{-9}\ s^{-1}$

D $mol^{-3}\ dm^9\ s^{-1}$

1 4 In an equilibrium system, the forward reaction is exothermic. Which of the following statements are correct?

[1 mark]

A The equilibrium will be unaffected by changes in pressure ⬭

B An increase in temperature decreases the proportion of products ⬭

C An increase in temperature decreases the proportion of reactants ⬭

D The rate of the forward reaction is unaffected by a change in temperature ⬭

1 5 Which combination is most likely to produce an alkene from a halogenoalkane?

[1 mark]

A Concentrated potassium hydroxide in ethanol, high temperature ⬭

B Concentrated aqueous potassium hydroxide, low temperature ⬭

C Dilute potassium hydroxide in ethanol, low temperature ⬭

D Dilute aqueous potassium hydroxide, high temperature ⬭

1 6 Which of the following will give no reaction with a mixture of potassium dichromate and sulfuric acid?

[1 mark]

A Propanone ⬭

B Propan-1-ol ⬭

C Propan-2-ol ⬭

D Propanal ⬭

1 7 What is the shape of a transition metal complex ion with a co-ordination number of 6?

[1 mark]

A Trigonal bipyramidal

B Tetrahedral

C Octahedral

D Square planar

1 8 Which of the following decreases across the Period Na–Ar?

[1 mark]

A Nuclear charge

B First ionisation energy

C Ionic radius

D Atomic radius

1 9 What is the volume occupied by 0.5 moles of carbon dioxide at 373 K and 100 kPa?

[1 mark]

A 30.6

B 22.4

C 15.3

D 24.0

2 0 An equilibrium mixture of the following:

$$2NO_{2(g)} \rightleftharpoons O_{2(g)} + 2NO_{(g)}$$

at 500 K and 1 atm, contains 0.96 mol of $NO_2(g)$, 0.04 mol of NO(g) and 0.02 mol of $O_2(g)$.

What is the value of K_p for this reaction, under these conditions?

[1 mark]

A 8.65×10^{-4} atm

B 3.39×10^{-5} atm

C 8.13×10^{-4} atm

D 8.13×10^{-4} atm

2 1 Which of these hydroxides is the least soluble in water?

[1 mark]

A $Mg(OH)_2$

B NaOH

C KOH

D $Ca(OH)_2$

2 2 Which equation represents the first electron affinity of chlorine?

[1 mark]

A $Cl_2(g) \rightarrow 2Cl(g)^+ + 2e^-$

B $Cl(g) + e^- \rightarrow Cl(g)^-$

C $Cl_2(g) + 2e^- \rightarrow 2Cl(g)^-$

D $Cl(g) \rightarrow Cl(g)^- + e^-$

2 3 Which chemical species can act as a nucleophile?

[1 mark]

A Br_2

B HBr

C Br•

D Br^-

2 4 What is the pH of a 1.50 mol dm^{-3} solution of ethanoic acid?

$(K_a = 1.7 \times 10^{-5})$

[1 mark]

A 2.30

B 4.50

C 5.05

D 3.35

2 5 Which reaction is spontaneous only at high temperatures?

[1 mark]

	ΔH^{θ}	$\Delta S^{\theta}_{system}$	
A	negative	positive	
B	positive	positive	
C	negative	negative	
D	positive	negative	

2 6 Element X has four isotopes: ^{82}X (12%), ^{83}X (12%), ^{84}X (50%), and ^{86}X (26%). What is the A_r of X?

[1 mark]

A 82.42 ⬭

B 84.16 ⬭

C 83.35 ⬭

D 85.20 ⬭

2 7 Which statement best describes an Arrehenius plot?

[1 mark]

A ln k against 1/T ⬭

B k against ln 1/T ⬭

C 1/T against k ⬭

D T against 1/ln$k$ ⬭

2 8 On a thin-layer chromatogram, the solvent front is at 28.7 cm and the spot produced by an amino acid is at 13.9 cm. What is the R_f value of the amino acid?

[1 mark]

A 2.06 ⬭

B 0.48 ⬭

C 0.20 ⬭

D 4.84 ⬭

2 9 Which is the correct list of successive ionisation energies for an element in Group 5 of the Periodic Table?

[1 mark]

A   1251, 2300, 3820, 5160, 6540, 9360 ⬭

B   578, 1820, 2750, 11600, 14800, 18400 ⬭

C   1010, 1900, 2910, 4960, 6270, 21269 ⬭

D   738, 1450, 7730, 10500, 13600, 18000 ⬭

3 0 What would you expect to observe if a solution containing silver nitrate and nitric acid is added to a solution of barium chloride?

[1 mark]

A   Effervescence ⬭

B   A white precipitate of silver chloride ⬭

C   A white precipitate of barium nitrate ⬭

D   No evidence of a reaction ⬭

3 1 What is the atom economy for the reduction of iron(III) oxide using carbon?

$$Fe_2O_3 + 3CO \rightarrow 2Fe + 3CO_2$$

[1 mark]

A   35% ⬭

B   68% ⬭

C   57% ⬭

D   46% ⬭

3 2 Which of the following molecules contains a chiral carbon atom?

[1 mark]

A   Propan-1-ol ⬭

B   Propan-1,2-diol ⬭

C   Propan-1,3-diol ⬭

D   Propan-1,2,3-triol ⬭

3 3 Nitration of 7.80 g of benzene produces 9.84 g of nitrobenzene. What is the percentage yield?

[1 mark]

A 100% ⬭

B 80% ⬭

C 60% ⬭

D 70% ⬭

3 4 Which of these species has a bond angle of 180°?

[1 mark]

A SO_2 ⬭

B CO_3^{2-} ⬭

C CO_2 ⬭

D NH_3 ⬭

3 5 What is the change in oxidation state of manganese when potassium manganate (VII) is oxidised by iron (II)?

[1 mark]

A from +7 to +3 ⬭

B from +7 to +6 ⬭

C from +7 to +5 ⬭

D from +7 to +2 ⬭

END OF QUESTIONS

Data booklet

The Periodic Table of the Elements

Key

relative atomic mass
atomic symbol
name
atomic (proton) number

Elements with atomic numbers 112–118 have been reported but not fully authenticated

1 (1)	2 (2)	(3)	(4)	(5)	(6)	(7)	(8)	(9)	(10)	(11)	(12)	3 (13)	4 (14)	5 (15)	6 (16)	7 (17)	0 (18)
1.0 **H** hydrogen 1																	4.0 **He** helium 2
6.9 **Li** lithium 3	9.0 **Be** beryllium 4											10.8 **B** boron 5	12.0 **C** carbon 6	14.0 **N** nitrogen 7	16.0 **O** oxygen 8	19.0 **F** fluorine 9	20.2 **Ne** neon 10
23.0 **Na** sodium 11	24.3 **Mg** magnesium 12											27.0 **Al** aluminium 13	28.1 **Si** silicon 14	31.0 **P** phosphorus 15	32.1 **S** sulfur 16	35.5 **Cl** chlorine 17	39.9 **Ar** argon 18
39.1 **K** potassium 19	40.1 **Ca** calcium 20	45.0 **Sc** scandium 21	47.9 **Ti** titanium 22	50.9 **V** vanadium 23	52.0 **Cr** chromium 24	54.9 **Mn** manganese 25	55.8 **Fe** iron 26	58.9 **Co** cobalt 27	58.7 **Ni** nickel 28	63.5 **Cu** copper 29	65.4 **Zn** zinc 30	69.7 **Ga** gallium 31	72.6 **Ge** germanium 32	74.9 **As** arsenic 33	79.0 **Se** selenium 34	79.9 **Br** bromine 35	83.8 **Kr** krypton 36
85.5 **Rb** rubidium 37	87.6 **Sr** strontium 38	88.9 **Y** yttrium 39	91.2 **Zr** zirconium 40	92.9 **Nb** niobium 41	96.0 **Mo** molybdenum 42	[98] **Tc** technetium 43	101.1 **Ru** ruthenium 44	102.9 **Rh** rhodium 45	106.4 **Pd** palladium 46	107.9 **Ag** silver 47	112.4 **Cd** cadmium 48	114.8 **In** indium 49	118.7 **Sn** tin 50	121.8 **Sb** antimony 51	127.6 **Te** tellurium 52	126.9 **I** iodine 53	131.3 **Xe** xenon 54
132.9 **Cs** caesium 55	137.3 **Ba** barium 56	138.9 **La*** lanthanum 57	178.5 **Hf** hafnium 72	180.9 **Ta** tantalum 73	183.8 **W** tungsten 74	186.2 **Re** rhenium 75	190.2 **Os** osmium 76	192.2 **Ir** iridium 77	195.1 **Pt** platinum 78	197.0 **Au** gold 79	200.6 **Hg** mercury 80	204.4 **Tl** thallium 81	207.2 **Pb** lead 82	209.0 **Bi** bismuth 83	[209] **Po** polonium 84	[210] **At** astatine 85	[222] **Rn** radon 86
[223] **Fr** francium 87	[226] **Ra** radium 88	[227] **Ac†** actinium 89	[267] **Rf** rutherfordium 104	[270] **Db** dubnium 105	[269] **Sg** seaborgium 106	[270] **Bh** bohrium 107	[270] **Hs** hassium 108	[278] **Mt** meitnerium 109	[281] **Ds** darmstadtium 110	[281] **Rg** roentgenium 111	[285] **Cn** copernicium 112	[286] **Nh** nihonium 113	[289] **Fl** flerovium 114	[289] **Mc** moscovium 115	[293] **Lv** livermorium 116	[294] **Ts** tennessine 117	[294] **Og** oganesson 118

* 58 – 71 Lanthanides

140.1 **Ce** cerium 58	140.9 **Pr** praseodymium 59	144.2 **Nd** neodymium 60	[145] **Pm** promethium 61	150.4 **Sm** samarium 62	152.0 **Eu** europium 63	157.3 **Gd** gadolinium 64	158.9 **Tb** terbium 65	162.5 **Dy** dysprosium 66	164.9 **Ho** holmium 67	167.3 **Er** erbium 68	168.9 **Tm** thulium 69	173.0 **Yb** ytterbium 70	175.0 **Lu** lutetium 71

† 90 – 103 Actinides

232.0 **Th** thorium 90	231.0 **Pa** protactinium 91	238.0 **U** uranium 92	[237] **Np** neptunium 93	[244] **Pu** plutonium 94	[243] **Am** americium 95	[247] **Cm** curium 96	[247] **Bk** berkelium 97	[251] **Cf** californium 98	[252] **Es** einsteinium 99	[257] **Fm** fermium 100	[258] **Md** mendelevium 101	[259] **No** nobelium 102	[262] **Lr** lawrencium 103

Chemistry Data Sheet

Table A
Infrared absorption data

Bond	Wavenumber/cm^{-1}
N—H (amines)	3300–3500
O—H (alcohols)	3230–3550
C—H	2850–3300
O—H (acids)	2500–3000
C≡N	2220–2660
C=O	1680–1750
C=C	1620–1680
C—O	1000–1300
C—C	750–1100

Table B
^{1}H NMR chemical shift data

Type of proton	δ/ppm
ROH	0.5–5.0
RCH$_3$	0.7–1.2
RNH$_2$	1.0–4.5
R$_2$CH$_2$	1.2–1.4
R$_3$CH	1.4–1.6
R—C(=O)—C—H	2.1–2.6
R—O—C—H	3.1–3.9
RCH$_2$Cl or Br	3.1–4.2
R—C(=O)—O—C—H	3.7–4.1
C=C—H	4.5–6.0
R—C(=O)—H	9.0–10.0
R—C(=O)—O—H	10.0–12.0

Table C
^{13}C NMR chemical shift data

Type of carbon		δ/ppm
—C—C—		5–40
R—C—Cl or Br		10–70
R—C(=O)—		20–50
R—C—N		25–60
—C—O—	alcohols, ethers or esters	50–90
C=C		90–150
R—C≡N		110–125
(benzene ring)		110–160
R—C=O	esters or acids	160–185
R—C=O	aldehydes or ketones	190–220

Phosphate and sugars

phosphate   glucose   2-deoxyribose

Bases

adenine   guanine   cytosine   thymine

Amino acids

alanine   aspartic acid   cysteine

lysine   phenylalanine   serine

Haem B

Answers

1. Isotopes have the same number and type of fundamental particles [1]
2. $X (g) + e^- \rightarrow X^+ (g) + 2e^-$ [1]
3. a) Mass number is the number of protons and neutrons in the nucleus of an atom. [1]

   b)

	Number of protons	Number of neutrons	Number of electrons
^6_3Li	3	3	3
$^7_3Li^+$	3	4	2

   [1 mark for each correct row]

4. a) To identify elements [1]

   To determine relative (molecular/atomic) mass [1]

   b) To interact with the electric field in order to be accelerated [1]

   To produce a current in the detector [1]

   c) i) $^{90}_{40}Zr$ [1 mark for correct mass number; 1 mark for correct atomic number]

   ii) $\dfrac{[(90 \times 51.5) + (11.2 \times 91) + (17.1 \times 92) + (17.4 \times 94) + (2.8 \times 96)]}{100}$ [1]

5. a) $(2 \times 1.673 \times 10^{-27}) + 0.911 \times 10^{-30}$ [1]

   $= 3.346\ 911 \times 10^{-27} = 3.35 \times 10^{-27}$ [1]

   b) $-1.602 \times 10^{-19}\,C$ [1]

   c) ^1_1H [1]

   99.98% [1]

1. Ca [1]
2. Oxygen [1]
3. a) i) $1s^2\,2s^2\,2p^6$ [1]

   ii) $1s^2\,2s^2\,2p^6\,3s^2\,3p^6\,3d^3$ [1]

   b) Usually the 4s completely fills before the 3d is filled. [1]

   But it is more stable/lower energy to have a completely filled 3d sub-level and a half filled 4s sub-orbital. [1]

   This is achieved by moving one electron from the 4s orbital to the 3d orbital. [1]

   c) Mg^{2+} [1]

   d) LiH [1]

4. a) The amount of energy required to remove an electron from an atom [1] in a gaseous state [1] to create a gaseous ion with a charge of 1^+ [1]

   b) $Cl(g) \rightarrow Cl^+(g) + e^-$ [1]

   c) i) Increases across a period [1]

   ii) Sulfur has paired electrons in its 3p sub-level [1]

   Phosphorus has an unpaired electron in its 3p sub-level [1]

   Pairing of electrons causes repulsion [1]

   Paired electrons are higher in energy and therefore easier to remove an electron from sulfur than from phosphorus [1]

   iii) Group 6 [1]

   Large jump in successive ionisation energy after 6th (between 6th and 7th) [1]

1. 1.74×10^{-2} [1]
2. $1.204 \times 10^{24}\,mol^{-1}$ [1]
3. a) $(2 \times 1.0) + (2 \times 16.0) = 34.0$ [1]

   b) $n = CV$ [1]

   $= 1.76 \times \left(\dfrac{25.00}{1000}\right)$ [1]

   $= 0.044$ moles [1]

   c) $0.050 \times \left(\dfrac{20.00}{1000}\right) \times 34.0$ [1]

   $= 0.034\,g$ [1]

4. a) $2HCl + Na_2CO_3 \rightarrow 2NaCl + H_2O + CO_2$

   [1 mark for balancing; 1 mark for correct products]

   b) $n = CV = 0.01 \times \left(\dfrac{22.70}{1000}\right)$ [1]

   $= 0.000\ 227$ moles [1]

   c) number of moles of acid $= 0.000\ 227 \times 2$

   $= 0.000\ 454$ [1]

   $\dfrac{(0.000\ 454 \times 1000)}{25.00} = 0.01816\,mol\,dm^{-3}$ [1]

   $1.8 \times 10^{-2}\,mol\,dm^{-3}$ [1]

1. $2.03\,dm^3$ [1]
2. Propene [1]
3. a) $C = \dfrac{83.7}{12.0}$ and $H = \dfrac{16.3}{1.0}$ [1]

   Whole number ratio C : H = 3 : 7 [1]

   Empirical formula $= C_3H_7$ [1]

   b) Relative mass of the empirical formula $= 43.0$ [1]

   $\dfrac{86.0}{43.0} = 2$ [1]

   Molecular formula $= C_6H_{14}$ [1]

4. Ideal gas equation: $pV = nRT$ [1]

   $103 \times 1000 = 103\ 000\,Pa$ [1]

   $n = \dfrac{pV}{RT} = \dfrac{103000 \times 127 \times 10^{-6}}{(8.31 \times 415)}$ [1]

   $= 2.51 \times 10^{-3}$ (mol) [1]

   $M_r = \dfrac{m}{n} = \dfrac{0.201}{2.51} \times 10^{-3} = 80.1$ [1]

Page 15

1. Hydration of ethene [1]
2. 32.83 g [1]
3. $12.0 + 3 \times 35.5 + 19.0 = 137.5$ and $121.8 + 3 \times 19.0 + 2 \times 79.9 + 12.0 + 4 \times 35.5 = 292.6$ [1]
   $\frac{137.5}{492.6} \times 100$ [1]
   $= 28\%$ [1]
4. a) $Ag^+ (aq) + Cl^- (aq) \rightarrow AgCl\ (s)$ [1]
   b) $Mg^{2+} (aq) + 2OH^- (aq) \rightarrow Mg(OH)_2 (s)$ [1]
   c) $MgCl_2 (aq) + Ca (s) \rightarrow CaCl_2 (aq) + Mg\ (s)$
   Correct equation [1]
   Correct state symbols [1]

Page 17

1. $[Ne]3s^1$ [1]
2. $Mg(NO_3)_2$ [1]
3. a) Electrostatic force of attraction [1]
   between oppositely charged ions [1]
   b) Lattice [1]
   c) Ag_2SO_4 [1]
4. a)
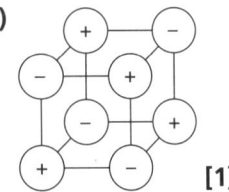
   [1]
   b) $1s^2 2s^2$ [1]
   c) Krypton (Kr) [1]

Page 19

1. Co-ordinate bond [1]
2. C_2H_4 [1]
3. a) Covalent [1]
   b) A shared pair of electrons [1]
   which are donated by one atom [1]
   c)
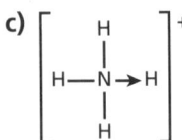
   For 4 covalent bonds N-H [1]
   For one bond as an arrow from N to H [1]
4. a) Type of bond: Covalent [1]
   How bond is formed: Shared pair of electrons [1]
   b) No lone pair of electrons (in the outer shell) [1]

Page 21

1. Diamond [1]
2. Aluminium [1]
3. a) Positive ions [1]
   (attract) delocalised electrons [1]
   b) Giant structure (accept metal crystal) [1]
   c) Lithium has the stronger bonding [1]
   Lithium ions are smaller / fewer electron shells [1]

Greater effective nuclear charge / less shielding from electrons [1]

4. a) $1s^2 2s^2 2p^6$ [1]
   b) Metallic [1]
   c)
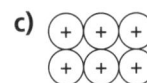
   OR
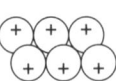
   OR
   $Na^+ Na^+ Na^+$
   $Na^+ Na^+ Na^+$
   For regular arrangement of particles [1]
   For $+/1+/+1$ in each particle [1]

Page 23

1. Silicon dioxide [1]
2. Molecular [1]
3. a) Requires a lot of energy to overcome ionic bonds [1]
   Many ionic bonds/(Giant) ionic lattice / lots of Mg^{2+} and Br^- ions [1]
   Strong (electrostatic) forces of attraction [1]
   between oppositely charged ions/ Mg^{2+} and Br^- ions [1]
   b) Ions [1] are free to move [1] and carry the charge [1]
4. a) Structure: covalent bonding [1]
   Bonding: molecular [1]
   b) Temperature measures the motion of the particles [1]
   When the substance is melting, the energy is being used to overcome the forces between the molecules [1] rather than increasing the motion of the particles [1]
   c)
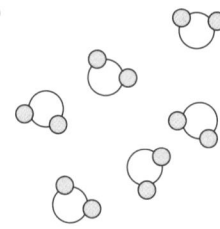
   For each molecule being $2 \times H$ (red circle) and $1 \times O$ (white circle) [1]
   For molecules being far apart (not touching) and random arrangement [1]

Page 25

1. Ammonium ion [1]
2. C(diamond) [1]
3. a) Octahedral [1]
   $90°$ [1]

b) i) Bent / V-shaped [1]

ii)

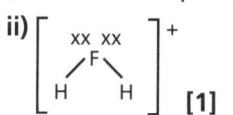 [1]

iii) Approximately 104.5° between the hydrogen atoms. [1]

4. Central element is O with six electrons and three covalent bonds, so nine electrons. [1]

But the ion has an overall charge of 1+, so actually eight electrons [1]

This is four electron pairs [1]

and therefore a tetrahedral shape [1]

with three bonding pairs and one lone pair of electrons. [1]

The bond angle would be about 107°. [1]

1. CH_3Br [1]

2. H–F [1]

3. a) Unsymmetrical electron distribution in a covalent bond [1]

between elements with different electronegativities [1]

b) There are permanent dipoles in the O–H bonds [1]

Water has a bent/V-shape [1]

This causes an uneven distribution of electrons in the whole molecule [1]

c)

Correct displayed formula of ammonia and water [1]

Correct hydrogen bonds from lone pair of electrons on O or N to hydrogen on the other molecule [1]

4. a) The ability of an atom to attract a pair of electrons / the electrons / electron density in a covalent bond [1]

b) Br is more electronegative than C [1]

So, Br is δ- and C is δ+ [1]

1. $2Na(s) + \frac{1}{2}O_2(g) \rightarrow Na_2O(s)$ [1]

2. Final temperature of the water [1]

3. a) 298 K [1]

100 kPa [1]

b) The energy change when one mole of a substance [1] is completely combusted [1] under standard conditions where all substances are in their standard state. [1]

c) $2C(s) + 3H_2(g) + 3\frac{1}{2}O_2(g) \rightarrow CH_3CH_2OH(l) + 3O_2(g)$

Balanced [1] with respect to one mole of ethanol produced [1]

Correct state symbols [1]

4. Temperature rise = 59.2 °C = 59.2 K and mass = 150 g [1]

$q = 150 \times 4.18 \times 59.2 = 37\,118.4$ J = 37.1184 kJ [1]

Heat change per mole = 37.1184 kJ ÷ 0.0300

$= 1237.28$ kJ mol^{-1} [1]

$\Delta_c H = -1237.28$ kJ mol^{-1} [1]

1. −1559.72 kJ mol^{-1} [1]

2. −283 kJ mol^{-1} [1]

3. a) The enthalpy change of a reaction is independent of the route taken. [1]

b) $\Delta_r H^{\ominus} = \Sigma \Delta_f H^{\ominus}$ (products) − $\Sigma \Delta_f H^{\ominus}$ (reactants) or correct Hess cycle [1]

$\Delta_r H^{\ominus} = -822 + (2 \times -272) = -278$ [1]

kJ mol^{-1} [1]

4. $\Delta_f H^{\ominus} = \Sigma \Delta_c H^{\ominus}$ (reactants) − $\Sigma \Delta_c H^{\ominus}$ (products) or correct Hess cycle [1]

$\Delta_f H^{\ominus} = (4 \times -394 + 4 \times -286) - (-2717)$

$= -3$ kJ mol^{-1} [1]

1. +2059 kJ mol^{-1} [1]

2. +431.8 [1]

3. a) 4(Xe–F) = 252 + (2 × 158) = 568 [1]

Xe–F = $\frac{568}{4}$ [1]

Xe–F = 142 [1]

kJ mol^{-1} [1]

b) Each specific bond will have a different bond enthalpy value in different molecules/environments. [1]

As the mean bond enthalpy is an average taken from a range of compounds, it will be different to a specific bond enthalpy value. [1]

4. a) The enthalpy change to break 1 mole of H–O/specific bonds [1] averaged over a range of compounds/molecules [1] where all species are in the gaseous state. [1]

b) $H_2(g) + \frac{1}{2}O_2(g) \rightarrow H_2O(g)$ or $\Delta H = (H-H) + \frac{1}{2}(O=O) - 2(H-O)$ or $\Delta H = \Sigma$bond energies of the reactants − Σbond energies of the products [1]

$= \frac{(436 + 496)}{(2 - 2 \times 464)}$ [1]

$= -244$ (kJ mol^{-1}) [1]

Page 35

1. The total number of particles present [1]
2. Average kinetic energy is constant at a fixed temperature. [1]
3. The reaction is exothermic/gives out heat [1]

   As concentration increases, the amount of heat given out increases/temperature increases more [1]

   More successful collisions in a given time / More particles have the activation energy [1]

   (An increase in temperature or more heat given out) increases the rate of a reaction [1]
4. a) Energy [1]

   b) Most probable/modal energy [1]

   c) There are no molecules with no energy [1]

   d) There is no maximum energy for molecules [1]

   e) Same volume under the curve [1]

   Curve displaced to the right, with lower peak [1]

Page 37

1. Adding a suitable catalyst [1]
2.

Rate of reaction	Yield of ammonia	Tick one box
Increases	Decreases	✓

[1]

3. a) As pressure was increased, the yield of C decreased [1]

   This is because the system opposes the change [1]

   Equilibrium shifts to the left as the reactants have fewer moles of gas [1]

   b) Lower yield at higher temperature [1]

   So, equilibrium shifted backwards/left direction to oppose the increase in temperature [1]

   This suggests that the forward reaction was exothermic [1]
4. a) Rate of forward reaction equals rate of reverse reaction [1]

   Concentrations of all substances are constant [1]

   b) No effect on position of equilibrium [1]

   but dynamic equilibrium is achieved faster [1]

   as the rate of the forward reaction and the rate of the reverse reaction are increased by the same amount [1]

Page 39

1. $mol^{-2} dm^6$ [1]
2. Increasing the temperature decreases the value of K_c [1]
3. a) All substances are in the same phase [1]

   b) $K_c = \dfrac{[C]^3}{[A][B]}$ [1]

   c) $mol\, dm^{-3}$ [1]

---

d) K_c is unchanged [1]

   Catalyst increases the rate of the forward and reverse reaction by the same amount [1]

   So the position of equilibrium is unchanged [1]

4. $K_c = \dfrac{[ClONO_2]}{[ClO][NO_2]}$ [1]

   Units = $mol^{-1} dm^3$ [1]

   For working out moles of species at equilibrium [1]

   For working out the concentration of each species at equilibrium [1]

	ClO	NO$_2$	ClONO$_2$
Starting amount (mol)	5	6	0
Amount at equilibrium (mol)	1	2	4
Change in amount (mol)	−4	−4	+4
Concentration (mol dm^{-3})	1 ÷ 2.5 = 0.4	2 ÷ 2.5 = 0.8	4 ÷ 2.5 = 1.6

$K_c = \dfrac{1.6}{0.4 \times 0.8}$ [1]

= 5 $mol^{-1} dm^3$ [1]

Page 41

1. Electron acceptor [1]
2. Fe_2O_3 [1]
3. a) NO = −2 [1]

   O$_2$ = 0 [1]

   NO$_2$ = +4 [1]

   b) The nitrogen in NO is being oxidised/increased oxidation number [1]

   The elemental oxygen is being reduced/decreased oxidation number [1]

   Both the oxidation and reduction are happening at the same time [1]

   c) O$_2$ [1]
4. a) $NaClO_3$ = +5 [1] and NaCl = −1 [1]

   b) Cl_2 is being reduced/gains electrons/electron acceptor/oxidising agent to form NaCl [1]

   Cl_2 is being oxidised/reducing agent to form $NaClO_3$ [1]

   This is an example of a disproportionation reaction [1]

   as the oxidation state of the same species/Cl_2 is increasing and decreasing [1]

Page 43

1. $\frac{1}{2}Br_2$ (l) → Br (g) [1]
2. LiI [1]
3. a) Enthalpy change or heat energy change when 1 mol of solid ionic compound/substance or 1 mol of ionic lattice [1]

   is formed from its gaseous ions [1]

b) $\Delta_{sol}H = \Delta_{Latt}H + \Delta_{hyd}H$ (calcium ions) + $2 \times \Delta_{hyd}H$ (chloride ions) **[1]**

$\Delta_{latt}H = -82 - 9 - (-1650 + 2 \times -364)$

$= +2295$ (kJ mol^{-1})$O_2 = 0$ **[1]**

$NO_2 = +4$ **[1]**

c) Beryllium ion is smaller than the calcium ion **[1]**

Therefore, it attracts the chloride ion more strongly / stronger ionic bonding **[1]**

4. a) $Ag(g) \rightarrow Ag^+(g) + e^-$ **[1]**

b) First electron affinity for chlorine **[1]**

c) Top line: $+ e^- + Cl(g)$ **[1]**

Second line from top: $+ e^- + \frac{1}{2}Cl_2(g)$ **[1]**

Bottom two lines: $+\frac{1}{2}Cl_2(g)$ **[1]**

Page 45

1. Lattice dissociation enthalpy is always a negative enthalpy change **[1]**

2. This is the value of enthalpy of lattice formation **[1]**

3. ΔH_1 = standard enthalpy of formation of CsCl **[1]**

ΔH_2 = standard enthalpy of atomisation of Cs **[1]**

ΔH_3 = first ionisation enthalpy of Cs **[1]**

ΔH_4 = standard enthalpy of atomisation of Cl **[1]**

b) It involves only bond making. **[1]**

c) $-(-364) - (+121) - (+376) - (+79) + (-433)$

$= -645$ **[1]**

kJ mol^{-1} **[1]**

4.

$Ba^{2+}(g) + 2e^{(-)} + 2Cl(g)$

BE of Cl_2 or $2\Delta Ha$ Chlorine **[1]** $2 \times EA$(chlorine) **[1]**
$Ba^{2+}(g) + 2e^{(-)} + Cl_2(g)$ $Ba^{2+}(g) + 2Cl^-(g)$

2nd IE(Ba) **[1]**
$Ba^+(g) + e^{(-)} + Cl_2(g)$

1st 1E (Ba) **[1]**   Lattice
$Ba(g) + Cl_2(g)$ (formation)
$\Delta Ha(Ba)$ or $\Delta Hsub(Ba)$ **[1]**   enthalpy
NOT $\Delta Huap$ Ba   or
$Ba(s) + Cl_2(g)$ energy
ΔH_f

$BaCl_2(s)$

$\Delta_a H$ barium + 1st $\Delta_{IE}H$ barium + 2nd $\Delta_{IE}H$ barium + 2$\Delta_a H$ for chlorine + 2$\Delta_{EA}H$ for chlorine + $\Delta_{LE}H$ − ΔH_f for barium chloride = 0 **[1]**

$+180 + 503 + 965 + 2 \times 122 + 2EA - 2056 + 859$

$= 0$ **[1]**

$EA = -\frac{695}{2} = -(347 \text{ to } 348)$ **[1]**

Page 47

1. 100 cm^3 of boiling saturated sodium chloride solution **[1]**

2. The reaction is feasible below a certain temperature **[1]**

3. a) At 0 K, particles are stationary **[1]** so have maximum order/no disorder/perfect order **[1]**

b) As temperature increases, particles vibrate/move **[1]** and disorder increases/order decreases **[1]**

c) Both changing state **[1]**

L_1 melting and L_2 boiling **[1]**

$L_1 < L_2$ bigger change in disorder for L_2 **[1]**

4. $\Delta H = (-201 + -242) - (-394)$ **[1]**

$\Delta H = -49$ kJ mol^{-1} **[1]**

$\Delta S = -180$ J K^{-1} mol^{-1} **[1]**

$\Delta G = \Delta H - T\Delta S$ **[1]**

$\Delta G = -49 - \left(890 \times -\frac{180}{1000}\right)$ **[1]**

$= 111$ kJ mol^{-1} **[1]**

Page 49

1. When the concentration of X doubles, the rate also doubles. **[1]**

2. mol^{-1} dm^3 s^{-1} **[1]**

3. a) 1 **[1]**

b) 0 **[1]**

c) Rate = k[CH$_3$I] **[1]**

d) s^{-1} **[1]**

4. $A = \dfrac{k}{e^{-Ea/RT}}$ **[1]**

$= \dfrac{(3.46 \times 10^{-8})}{e^{-96\,200/(8.31 \times 298)}}$ **[1]**

$= 2.57 \times 10^9$ **[1]**

s^{-1} **[1]**

Page 51

1. kPa^{-1} **[1]**

2. Changing the temperature changes the value of K_p. **[1]**

3. a) $K_p = \dfrac{P_{PCl_5}}{P_{PCl_3} \times P_{Cl_2}}$ **[1]**

b) Pa^{-1} (accept kPa^{-1} or MPa^{-1}) **[1]**

c) K_p decreases **[1]**

(Le Chatelier's Principle states) the reaction will shift to oppose the change **[1]**

move in the backwards/ favour endothermic direction / equilibrium position shifts to the left **[1]**

4. $K_p = \dfrac{P_{SO_2} \times P_{Cl_2}}{P_{SO_2Cl_2}}$ **[1]**

Total moles of gas in mixture = 0.25 + 0.75 + 0.75 = 1.75 **[1]**

p = total pressure × mol fraction **[1]**

Partial of SO$_2$Cl$_2$: $125 \times \frac{0.25}{1.75} = 17.9$ kPa **[1]**

Partial pressure of Cl$_2$: $125 \times \frac{0.75}{1.75} = 53.6$ kPa **[1]**

$K_p = 53.6 \times \frac{53.6}{17.9}$

$= 161$ **[1]**

kPa **[1]**

1. +1.62 V [1]
2. A salt bridge completes the circuit. [1]
3. a) 100 kPa [1]
   298 K [1]
   All solutions 1 mol dm^{-3} [1]
   b) Standard hydrogen electrode [1]
   c) $Fe^{3+}(aq) + 2e^- \rightleftharpoons Fe^{2+}(aq)$ [1]
   d) Hydrogen atoms are oxidised/lose an electron [1]
      The electron is passed to the iron(III) ions [1]
      Iron(III) ions are reduced/gain an electron [1]
4. a) $Cu|Cu^{2+}||Ag^+|Ag$
      Correct species as anode and cathode [1]
      With salt bridge [1]
   b) $Cu + 2Ag^+ \rightarrow Cu^{2+} + 2Ag$
      Correct species [1]
      Correct balancing [1]
   c) $E^{\varnothing}_{cell} = E^{\varnothing}_{Right} - E^{\varnothing}_{Right}$ [1]
      $E^{\varnothing}_{cell} = +0.80 - +0.34 = +0.46$ V [1]

1. Co [1]
2. +1.40 V [1]
3. a) $Li^+(aq) + MnO_2(s) + e^- \rightarrow LiMnO_2(s)$
      Correct equation [1]
      State symbols [1]
   b) $Li(s) \rightarrow Li^+(aq) + e^-$
      Correct equation [1]
      State symbols [1]
   c) −0.13 (V) [1]
4. a) Positive electrode: $O_2 + 2H_2O + 4e^- \rightarrow 4OH^-$ [1]
      Negative electrode $H_2 + 2OH^- \rightarrow 2H_2O + 2e^-$ [1]
   b) $2H_2 + O_2 \rightarrow 2H_2O$ [1]
   c) Fuel cell does not give out pollutants such as NO_x or CO_2 or SO_2 or C or CO or C_xH_y or unburnt hydrocarbons / water is the only product [1]

1. H_3O^+ [1]
2. 1.40 [1]
3. a) $K_c = \frac{[H^+][OH^-]}{[H_2O]}$ [1]
   b) $[H_2O]$ almost constant/very large/equilibrium position very towards the left. [1]
   c) Forward reaction is endothermic therefore position of equilibrium shifts to the right when temperature increases [1]

This increases $[H^+]$ [1]
and so the pH increases [1]
   d) $[H^+] = [OH^-]$ at all temperatures [1]
4. a) Proton acceptor [1]
   b) $CH_3COOH + HNO_3 \rightarrow CH_3COOH_2^+ + NO_3$ [1]
   c) $pH = -\log10[H^+]$ [1]
      $[H^+] = 1 \times 10^{-1.35} = 0.045$ mol dm^{-3} [1]
      $[H^+] = [HNO_3] = 0.045$ mol dm^{-3} [1]

1. 2.98 [1]
2. Hydrocyanic acid is a weaker acid than nitrous acid. [1]
3. a) $K_a = \frac{[CH_3COO^-][H^+]}{[CH_3COOH]}$ or $\frac{[CH_3COO^-][H_3O^+]}{[CH_3COOH]}$ [1]
   b) $[H^+] = 10^{-2.69} = 2.042 \times 10^{-3}$ mol dm^{-3} [1]
      $[CH_3COOH] = \frac{[H^+]^2}{K_a}$ [1]
      $= \frac{(2.042 \times 10^{-3})^2}{1.75 \times 10^{-5}}$ [1]
      $= 0.238$ mol dm^{-3} [1]
   c) Chloroethanoic acid is a stronger acid than ethanoic acid [1] because K_a is a lower value [1] which means the position of equilibrium is more to the right / higher concentration of products / greater ionisation / greater dissociation [1]
4. $K_a = \frac{[H^+][X^-]}{[HX]}$ [1]
   $\approx K_a = \frac{[H^+]^2}{[HX]} = \frac{(4.57 \times 10^{-3})^2}{0.150}$ [1]
   $= 1.41 \times 10^{-4}$ mol dm^{-3} [1]
   $pK_a = -\log10K_a$ [1]
   $= 3.86$ [1]

1. Bromophenol blue (pH range 3.0–4.6) [1]
2. $CH_3CH_2COOH + KOH$ [1]
3. a) Volume at half equivalence = 9.75 cm^3 [1]
      pH = 4.9 [1]
      $K_a (= 10^{-pH}) = 10^{-4.9} = 1.26 \times 10^{-5}$ [1]
      mol dm^{-3} [1]
   b) The equivalence point is above pH 7. [1]
      Indicator would not change colour in suitable pH range. [1]
      Therefore, an indicator shouldn't be used for this titration as the end-point would not be a good approximation of the equivalence point. [1]
4. a) A strong acid, e.g. HCl, HNO_3 or H_2SO_4 [1]
      A weak base, e.g. NH_3, NH_4OH [1]
   b) Repeat the experiment with each indicator [1]
      Select the indicator that changes colour rapidly when the pH changes from about 7 to 4 [1]

1. Ethanoic acid and sodium ethanoate [1]
2. B [1]
3. a) Small addition of acid [1]
   Small addition of alkali/base [1]
   On dilution (with water) [1]
   b) ratio
   $\frac{[HX]}{[X^-]}$ [1]
   c) Addition of acid, increase in $[H^+]$, equilibrium moves to the left (reacts with carbonate ions) [1]
   Addition of alkali causes a reaction between OH^- and H^+ [1] and so a decrease in $[H^+]$; equilibrium shifts to the right (to replace H^+) [1]
   $[H^+]$ remains almost constant [1]
4. $[H^+] = 10^{-4.5} = 3.16 \times 10^{-5}$ mol dm^{-3} [1]
   $[C_2H_5COO^-] = \dfrac{[C_2H_5COOH]K_a}{[H^+]}$ [1]
   $[C_2H_5COO^-] = 0.1068$ mol dm^{-3} [1]
   M_r sodium propanoate = 96 [1]
   Amount of sodium propanoate = 0.1068×0.5
   $= 0.0534$ mol [1]
   Mass of sodium propanoate = 0.0534×96
   $= 5.13$ g [1]

1. All Period 3 elements are in the same block. [1]
2. f-block [1]
3. a) First ionisation energy increases across the period [1] as shielding is constant / atomic radius decreases [1] effective nuclear charge increases [1] the outer-shell electrons need more energy to be removed [1]
   b) Outer electron in 3p sub-shell [1]
   3p higher in energy / slightly more shielded than 3s / slightly further away than 3s [1]
4. a) Silicon (Si) [1]
   b) S_8 molecules are bigger than P_4 molecules [1]
   Larger molecules are more easily polarisable [1]
   so stronger dipole-dipole interactions between S_8 than between P_4 [1]

1. Magnesium is a reducing agent in the production of titanium. [1]
2. Forms a highly soluble hydroxide [1]
3. a) s-block [1]
   Highest occupied energy level is an s-orbital [1]
   b) $Mg\ (s) + H_2O\ (g) \rightarrow MgO\ (s) + H_2\ (g)$ [1]
   c) Bright/white flame/light [1]
   White/grey ash/powder/smoke [1]
   d) Mg is used in extracting Ti from $TiCl_4$ [1]
   $Mg(OH)_2$ is used as a laxative/medically [1]

4. a) Alkaline earth metals [1]
   b) It decreases [1]
   c) Magnesium/Mg [1] has a different lattice/crystal/structure to the other Group 2 metals [1]

1. F_2 [1]
2. The first ionisation energy of the element decreases. [1]
3. a) p-block [1]
   Highest occupied energy level is a p-orbital [1]
   b) Toxic/poisonous/too much chlorine causes death [1]
   c) $Cl_2 + H_2O \rightleftharpoons HCl + HClO$ [1]
   Cl_2 gains an electron to form Cl^- and Cl_2 loses an electron to form ClO^- [1]
   Both oxidation and reduction are happening to the same species [1]
4. a) Add dilute ammonia solution to the precipitate. [1]
   If the precipitate dissolves, chloride ions are present. [1]
   If the precipitate does not dissolve, bromide ions are present. [1]
   b) Add chlorine water. [1]
   If no visible change, chloride ions are present. [1]
   If an orange-brown solution forms, bromide ions are present. [1]

1. SO_3 [1]
2. Al_2O_3 [1]
3. a) Both SO_3 and SO_2 have simple molecular structures and no bonds are broken when they are melted; just dipole–dipole forces between molecules are overcome. [1]
   SO_3 is larger than SO_2 so more easily polarisable [1] and has stronger dipole–dipole forces. [1]
   b) $SO_2 + H_2O \rightarrow H_2SO_3$ [1]
   c)

   [1]
   d) $MgO + H_2SO_4 \rightarrow MgSO_4 + H_2O$ [1]
4. a) $P_4O_{10} + 6H_2O \rightarrow 4H_3PO_4$ [1]
   pH between 1 and 2 [1]
   b) SiO_2 is macromolecular / giant covalent / giant molecule. [1]
   Many strong covalent bonds between atoms need to be broken. [1]
   P_4O_{10} has a simple molecular structure. [1]

Only weak dipole-induced dipole forces between molecules need to be overcome. [1] More energy is needed to break strong bonds than overcome weak dipole-induced dipole forces. [1]

1. [Ar] $3d^1 4s^2$ [1]
2. NH_4^- [1]
3. a) Dative/co-ordinate bond [1]
      Both electrons from the oxygen/ligand [1] are donated to the central metal ion. [1]
   b) Oxidation state of copper = 2+ [1]
      Co-ordination number = 6 [1]
   c) (is a transition metal because) it has a stable ion with an incomplete d sub-shell [1]
      (is a d-block element because) it has highest energy electron in the d sub-shell [1]
4. a) [Ar] [1]
      The stable ion doesn't have any electrons in the d sub-shell. [1]
   b) Transition metal atom or ion bonded to one or more ligands [1] by co-ordinate / dative (covalent) bonds / donation of an electron pair. [1]

1. Ligand substitution [1]
2. $\Delta S > 0$ J mol^{-1} [1]
3. a) $[Fe(H_2O)_6]^{2+}$ [1]
   b) $[Fe(H_2O)_6]^{3+} + 4Cl^- \rightarrow [FeCl_4]^- + 6H_2O$ [1]
   c) Cl^- is a bigger than H_2O [1]
      Only 4 Cl^- can fit around the metal ion [1]
4. a) Two atoms that each donate a lone pair (of electrons) / co-ordinate bonds / dative bond [1]
   b) $[Cu(H_2O)_6]^{2+} + 3NH_2CH_2CH_2NH_2 \rightarrow [Cu(NH_2CH_2CH_2NH_2)_3]^{2+} + 6H_2O$ [1]
   c) There is an increase in the number of particles / the reaction goes from 4 moles to 7 moles [1]
      Disorder/entropy increases / ΔS is positive [1]
      ΔG negative [1]

1. $[Pt(NH_3)_2Cl_2]$ [1]
2. Cu^{2+} [1]
3. a) $[Co(H_2O)_6]^{2+}$ [1]
      Octahedral [1]
   b) Change in ligands and co-ordination number [1] caused ΔE to change/decrease. [1]
      In $[Co(H_2O)_6]^{2+}$ electrons absorbed blue/violet light, leaving the complementary colour pink to be visible. [1]

$[CoCl_4]^{2-}$ appears blue so the complementary colours of red/orange/yellow must have been absorbed. [1]
As red/orange/yellow are lower in energy than blue/violet, ΔE in $[Co(H_2O)_6]^{2+}$ is higher than in $[CoCl_4]^{2-}$. [1]
   c) 90° [1]
4. Calibrate a colorimeter / produce a calibration curve [1] by using the colorimeter with solutions of copper-EDTA complex of known concentrations. [1]
   Add excess EDTA salt to the sample. [1]
   Any one from:
   If using a calibrated colorimeter: measure the concentration of the sample on the colorimeter [1]
   If using a calibration curve: measure the transmittance of light and compare to the calibration curve. [1]

1. $FeSO_4$ [1]
2. VO^{2+} [1]
3. a) $[Ag(NH_3)_2]^+$ [1]
   b) Chemical test to distinguish between ketones and aldehydes [1]
4. $2MnO_4^- + 6H^+ + 5H_2O_2 \rightarrow 2Mn^{2+} + 8H_2O + 5O_2$ [1]
   Amount of $MnO_4^- = \dfrac{0.020 \times 35.75}{1000}$
   $= 7.15 \times 10^{-4}$ mol [1]
   Amount of $H_2O_2 = 7.15 \times 10^{-4} \times \dfrac{5}{2}$
   $= 1.786 \times 10^{-3}$ mol [1]
   $[H_2O_2]$ 5.00% sample $= \dfrac{1.786 \times 10^{-3}}{25 \times 10^{-3}}$
   $= 0.0714$ mol dm^{-3} [1]
   $[H_2O_2]$ original solution $= 0.0714 \times 20 = 1.43$ [1] mol dm^{-3} [1]

1. V_2O_5 [1]
2. Manganate(VII) ions and ethanoate ions [1]
3. a) Fe [1]
   b) A substance that is in the same phase as the reactants [1] which lowers the activation energy [1] by providing an alternative reaction pathway [1] which increases the rate of reaction. [1]
4. a) Any one from Mn^{2+}; Mn^{3+} [1]
   b) Mn can exist in variable oxidation states [1]
      Provide an alternative reaction pathway with lower E_a is because oppositely charged ions attract [1]
      Mn^{3+} reduced to Mn^{2+} by $C_2O_4^{2-}$ [1]
      Mn^{2+} oxidised to Mn^{3+} by MnO_4^- [1]

1. A white precipitate of aluminium(III) carbonate and bubbles of carbon dioxide gas [1]
2. $CuCl_2$ [1]
3. a) Green [1]
   b) $[Fe(H_2O)_6]^{2+} + 2NH_3 \rightarrow Fe(H_2O)_4(OH)_2 + 2NH_4^+$ [1]
   Green precipitate forms [1]
   No further change on excess ammonia [1]
   c) Oxidation [1]
   Results in a colour change (to brown) [1]
   Or, balanced equation: $Fe(H_2O)_4(OH)_2$ (s) $\rightarrow$ $Fe(H_2O)_3(OH)_3$ (s) + H^+ (aq) + e^- [2]
4. Reaction 1: Dilute ammonia (NH_3) [1]
   $[Cu(H_2O)_6]^{2+} + 2NH_3 \rightarrow [Cu(H_2O)_4(OH)_2] + 2NH_4^+$
   Or (solution) / NaOH [1]
   $[Cu(H_2O)_6]^{2+} + 2OH^- \rightarrow [Cu(H_2O)_4(OH)_2] + 2H_2O$ [1]
   Reaction 2: Concentrated or excess ammonia [1]
   $[Cu(H_2O)_4(OH)_2] + 4NH_3 \rightarrow [Cu(H_2O)_2(NH_3)_4]^{2+} + 2H_2O + 2OH^-$ [1]
   Reaction 3: Na_2CO_3 / any identified soluble carbonate [1]
   $[Cu(H_2O)_6]^{2+} + CO_3^{2-} \rightarrow CuCO_3 + 6H_2O$
   Or $NaHCO_3$ [1]
   $[Cu(H_2O)_6]^{2+} + HCO_3^- \rightarrow CuCO_3 + 6H_2O + H^+$ [1]

1. 2-methylpropan-1-ol [1]
2. Structural [1]
3. a) C_nH_{2n+2} [1]
   b)
   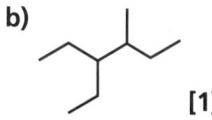
   [1]
   c) C_9H_{20} [1]
4. a) C_6H_{12} [1]
   b) C=C [1]
   c) i) 2,3-dichlorohexane [1]
   ii)
   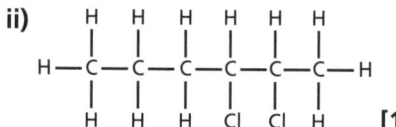
   [1]

1. $Cl_2 \rightarrow 2Cl^\bullet$ [1]
2. The movement of two electrons [1]
3. a) Free radical [1]
   Substitution [1]
   b) UV light [1]

   c) Br——Br $\rightarrow 2Br^\bullet$

   Single-headed arrows, one to each bromine atom originating from the centre of the bond [1]
   Two bromine radicals being formed [1]

4. 1 mark for each curly arrow correctly drawn [3]

1. 4-methylpent-2-ene [1]
2. 3 [1]
3. a)
   [1]
   b) Propanal [1]
   c) Has free rotation around all bonds [1]
   as doesn't contain C=C [1]
4. Use Cahn–Ingold–Prelog priority rules. [1]
   Focus on the C=C and the groups that come off each carbon atom to assess their priority. [1]
   C has an atomic mass of 12 and is priority 1; H has an atomic mass of 1 and is priority 2. [1]
   If the two priority 1 groups are on the same side of the molecule, it is the Z isomer; if they are on opposite sides, it is the E isomer. [1]
   The first isomer is therefore Z and the second isomer is E. [1]

1. A mixture of hydrocarbons with similar boiling points [1]
2. Alkanes contain the C=C functional group [1]
3. a) Fractional distillation [1]
   b) Thermal cracking [1]
   to make ethene/alkenes [1]
   c) Naphtha is a mixture. [1]
   Each substance in the mixture has a similar boiling point. [1]
4. a) Contains only single (covalent) bonds [1]
   b) i) Crude oil [1]
   ii) Fractional distillation [1]
   c) i) $C_{16}H_{34} \rightarrow C_8H_{16} + C_4H_{10} + C_4H_8$ [1]
   ii) High temperature and slightly raised pressure [1]
   with a zeolite catalyst [1]
   d) Smaller chain molecules are in more demand. [1]
   Smaller chain molecules have higher value. [1]

Page 93

1. Sulfur dioxide can be removed from car exhausts using a catalytic converter [1]

2. Free radical substitution [1]

3. a) Ultraviolet (UV) [1]

   b) Reaction is uncontrolled/many different radicals are produced [1]

   In the termination step, all the combinations of the many different radicals can join to make the products [1]

   c) $CH_3^\bullet + Cl^\bullet \rightarrow CH_3Cl$ [1]

   d) Fractional distillation [1]

4. a) $C_9H_{20} + 14O_2 \rightarrow 9CO_2 + 10H_2O$

   For C_9H_{20} [1]

   Correct products and balancing [1]

   b) i) Acid rain (accept respiratory problems/ smog/production of ground level ozone) [1]

   ii) Catalytic converters [1] contain rhodium/ platinum [1] which reduce NO_x into its elements/ $NO_x \rightarrow N_x + O_x$ [1]

Page 95

1. Nucleophilic substitution [1]

2. Cl_2 [1]

3. a) NaOH ionises [1]

   to make OH– [1]

   which can form a bond by donating a pair of electrons [1]

   b)

   H H H H   H H H H
   | | | | | | | |
   H—C—C—C—C—Br → H—C—C—C—C—O—H
   | | | | ↑ | | | |
   H H H H   H H H H
        :ÖH

   For each curly arrow [2]

   For displayed formula (every atom and every bond) of propan-1-ol [1]

   c) But-1-ene [1]

4. a) C_6H_{12} [1]

   b)

   HÖ:—↗

   H Br H H H   H   CH₂CH₂CH₃
   | | | | | \   /
   H—C—C—C—C—C—H   →   C=C
   | | | | | /   \
   H C H H H   H   CH₃
   | /\
   H H H
   2-methylpent-1-ene [1]

   Correct curly arrows [1]

   Correct structure of product [1]

Page 97 (right column top)

   H ⤹Br H H H   CH₃   H
   | | | | | \   /
   H—C—C—C—C—C—H   →   C=C
   | | | | | /   \
   H C H H H   CH₃   CH₂CH₃
   | /\   2-methylpent-2-ene [1]
   H H H :ÖH

   Correct curly arrows [1]

   Correct structure of product [1]

   c) The C–X is polar [1]

   giving the C a slightly positive charge [1]

Page 97

1. Ozone absorbs wavelengths of UV radiation. [1]

2. HFCs can catalyse the formation of ozone. [1]

3. a) $CClF_3 \rightarrow {}^\bullet CF_3 + Cl^\bullet$ [1]

   Chlorine radical/chlorine atom is the catalyst [1]

   b) $Cl^\bullet + O_3 \rightarrow ClO^\bullet + O_2$ [1]

   $ClO^\bullet + O_3 \rightarrow 2O_2 + Cl^\bullet$ [1]

4. a) $Br_2 \rightarrow 2Br^\bullet$ [1]

   UV light [1]

   b) C–Br breaks more readily / is weaker / has lower bond enthalpy than C–Cl [1]

   c) $Br^\bullet + O_3 \longrightarrow BrO^\bullet + O_2$ [1]

   $BrO^\bullet + O_3 \longrightarrow Br^\bullet + 2O_2$ [1]

   The bromine atom/radical does not appear in the overall equation [1]

   as it is regenerated and unchanged at the end [1]

   because it provided an alternative route/ mechanism with a lower activation energy [1]

Page 99

1. It is likely to undergo nucleophilic addition. [1]

2. Ethene forms a planar molecule. [1]

3. a) B [1]

   b) No rotation around C=C/functional group [1]

   Two different groups on each carbon atom of the C=C [1]

   c) Use bromine water [1]

   Shake with organic sample [1]

   Decolourises if unsaturated [1]

4. a) They are (organic substances that are) unsaturated [1]

   They are hydrocarbons [1]

   b)

   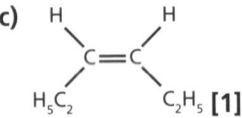 [1]

   c)

   H   H
    \   /
     C=C
    /   \
   H₅C₂   C₂H₅ [1]

Page 101

1. 3-bromo-3-methylpentane [1]

2. Sulfuric acid [1]

3. a) Electrophilic addition [1]
    b) Major:

    $$H_3C-\underset{\underset{\underset{SO_3H}{|}}{\underset{O}{|}}{\overset{\overset{CH_3}{|}}{C}}-\underset{\overset{H}{|}}{C}-CH_3$$ [1]

    Minor:

    $$H_3C-\underset{\underset{H}{|}}{\overset{\overset{CH_3}{|}}{C}}-\underset{\underset{\underset{SO_3H}{|}}{\underset{O}{|}}{\overset{\overset{H}{|}}{C}}-CH_3$$ [1]

    Correct labelling of major and minor. [1]
    c) Asymmetric alkene [1]
    The major product is formed via a tertiary carbocation intermediate and the minor product is formed via a secondary carbocation intermediate [1]
    The tertiary carbocation is more stable than the secondary carbocation [1]
4. a) Electrophilic addition [1]
    b)

    2-methyl-2-butene

    More stable carbocation → 2-bromo-2-methylbutane (major)

    Less stable carbocation → 2-bromo-3-methylbutane (minor)

    Initial step of mechanism [1]
    One mark for second step for each product [2]
    Determining the minor and major products [1]
    One mark for each name [2]

Page 103

1. It has a higher melting point than propene. [1]
2. Poly(buten-1-ol) [1]
3. a) Addition polymerisation [1]
    b)

    $$-\underset{\underset{H}{|}}{\overset{\overset{H}{|}}{C}}-\underset{\underset{H}{|}}{\overset{\overset{CH_3}{|}}{C}}-$$ [1]

    c) Bonding: Covalent bonds [1]
    made from shared pairs of electrons between the atoms [1]

Structure: Long-chain molecule [1]
induced dipole-dipole forces between polymer chains [1]

4. a) Poly(chlorotrifluoroethene) [1]
    b)

    $$\underset{\underset{F}{|}}{\overset{\overset{F}{|}}{C}}=\underset{\underset{Cl}{|}}{\overset{\overset{F}{|}}{C}}$$ [1]

    c) The melting point of Halon is higher than polyethene. [1]
    Polyethene has induced dipole-dipole forces between the polymer chains. [1]
    Halon has induced dipole-dipole forces and permanent dipole-dipole forces between the polymer chains. [1] There are stronger forces between Halon chains than between poly(ethene) chains and so more energy is needed to overcome them, resulting in a higher melting point. [1]

Page 105

1. The reaction is catalysed by zymase. [1]
2. Bioethanol is made from anaerobic fermentation with yeast. [1]
3. a) Fermentation [1]
    b) No net emissions of carbon dioxide to the atmosphere [1]
    c) $6CO_2 + 6H_2O \rightarrow C_6H_{12}O_6 + 6O_2$ [1]
    $C_6H_{12}O_6 \rightarrow 2C_2H_5OH + 2CO_2$ [1]
    $2C_2H_5OH + 6O_2 \rightarrow 4CO_2 + 6H_2O$ [1]
    6 moles of carbon removed by the plant during growing, 2 moles of carbon released to make the biofuel and 4 moles of carbon released to make the biofuel, so no net emissions. [1]
4. a) Fractional distillation [1] of crude oil [1] followed by steam cracking of the long-chain hydrocarbons. [1]
    b)

    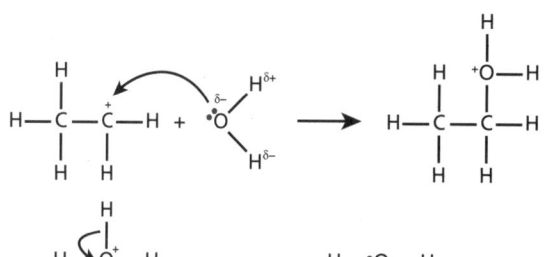

    One mark for each correct curly arrow [3]
    c) Concentrated phosphoric acid [1]

1. Concentrated sulfuric acid [1]
2. Blue solution turns brick red [1]
3. a) Heat [1] under reflux [1]
   b) Ethanedioic acid [1]
   c) Primary alcohol [1]
   The -OH/functional group/hydroxyl group is on the end of the carbon chain/first carbon atom. [1]
4. a) Reagent: acidified potassium dichromate [1]
   Solution of pentan-2-ol would change from orange to green. [1]
   Solution of 2-methylbutan-2-ol would have no observable change. [1]
   b) i) Pent-1-ene [1]
   Pent-2-ene [1]
   ii) Excess concentrated sulfuric/phosphoric acid [1]
   Aluminium oxide catalyst [1]
   Heat/vapours [1]

1. Acidified potassium dichromate [1]
2. $CH_2CHCH_2CH_2CH_2CH_3$ [1]
3. a) Carboxylic acid/–COOH [1]
   b) Any value between pH 3 and 6 [1]
   c) Butanoic acid [1]
   2-methylpropanoic acid [1]
4. a) Same molecular formula but a different functional group [1]
   Functional group isomers [1]
   b) Either:
   Fehling's solution [1]
   Warm [1]
   Substance B (hexanal) would cause the blue solution to form a brick red precipitate. [1]
   No observable change with substance A (hexan-2-one). [1]
   Or:
   Tollens' reagent [1]
   Warm [1]
   Substance B (hexanal) would produce a silver precipitate/silver mirror on the inside of the test tube. [1]
   No observable change with substance A (hexan-2-one). [1]
   Or:
   Acidified potassium dichromate [1]
   Warm [1]
   For substance B (hexanal), the orange solution would turn green. [1]
   No observable change with substance A (hexan-2-one). [1]

1. C_2H_4O [1]
2. CH_3O^+ [1]
3. a) Any one from: by definition; the standard; the reference [1]
   b) 60.05843 [1]
   60.0584 [1]
   c) Owing to the presence of ^{13}C [1]
   in the molecular ion. [1]
4. a) The molecule with:
   one/an electron knocked off/lost (or a single positive charge) [1]
   Or:
   the ion with m/z = M_r [1]
   b) Cl has two isotopes. [1]
   c) $CH_3 \overset{+}{C} = O$ [1]

1. O_2 [1]
2. Hexane [1]
3. a) C=O [1]
   b) Ester / $-\overset{\overset{O}{\|}}{C}-O^-$ [1]
   c) Compare the fingerprint region/infrared spectrum below 1500 cm^{-1} [1]
   to known spectra to find a match [1]
4. a) Butanone [1]
   b) i)
   Functional group is –OH [1]
   Correct structure [1]
   ii) Cyclobutanol [1]

1. A racemic mixture of 1,2-dichloropropene rotates polarised light. [1]
2. Dehydration of butan-2-ol by heating with concentrated sulfuric acid [1]
3. a)
   Structure [1]
   Identification of the chiral carbon [1]
   b) (Each of a pair of molecules) that are non-superimposable [1] mirror images of each other. [1]
   c) Equal proportion of both enantiomers [1] which rotate the plane of polarised light in opposite directions, [1] cancelling out each other / no overall effect. [1]

4. a) Molecules with same structure / structural formula **[1]** but with bonds/atoms/ groups arranged differently in space. **[1]**
    b) Plane of polarised light **[1]**
       rotates equally in opposite directions. **[1]**

Page 117

1. KCN and dilute H_2SO_4 **[1]**
2. 2,3-dimethylbutan-1-ol **[1]**
3.
    a) For each correct curly arrow **[2]**
       Correct intermediate structure **[1]**
    b) Nucleophilic addition **[1]**
    c) Hydroxynitrile **[1]**
       Contains a cyano (–CN) and a hydroxy (–OH) group **[1]** attached to the same carbon atom. **[1]**
4. a) Structural **[1]**
    b) i) T **[1]**
       ii) Silver mirror **[1]**
    c) i) T **[1]**
       ii) Orange solution turns green **[1]**
    d) T and S **[1]**

Page 119

1. Ethanol and concentrated sulfuric acid **[1]**
2. Vitamin C is a reductant. **[1]**
3. a) $3CH_3(CH_2)_{14}COOH$ **[1]**
       $CH_2(OH)CH(OH)CH_2OH$ (accept $C_3H_8O_3$) **[1]**
    b) Catalyst **[1]**
    c) Biofuel **[1]**
4. a) $CH_3CH_2CH_2COOH + CH_3CH_2OH \rightarrow$ $CH_3CH_2CH_2COOCH_2CH_3 + H_2O$
       Correct acid **[1]**
       Correct alcohol **[1]**
       Correct products **[1]**
    b) H_2SO_4 / HCl / H_3PO_4 **[1]**
       Heat **[1]**
       under reflux **[1]**
    c) i) –COO / R″—C(=O)—O—R′ **[1]**
       ii) Ester **[1]**

Page 121

1. Primary alcohol **[1]**
2. Nucleophilic addition-elimination **[1]**
3. a) i) $CH_3COOC_6H_4COOH + HCl$ **[1]**
       ii) $CH_3COOC_6H_4COOH + CH_3COOH_3$ **[1]**
          $CH_2(OH)CH(OH)CH_2OH$ (accept $C_3H_8O_3$) **[1]**
    b) Ethanoic anhydride is cheap compared to

ethanoyl chloride. **[1]**
       Safer as less corrosive than ethanoyl chloride or HCl evolved / reaction less violent or vigorous / less exothermic / more easily controlled. **[1]**
       Less vulnerable to hydrolysis. **[1]**
4. a) Methyl propanoate **[1]**
    b)
       For arrow and lone pair **[1]**
       For arrow addition-elimination **[1]**
       For structure **[1]**
       For three arrows **[1]**
       Name of mechanism: Nucleophilic addition-elimination **[1]**

Page 123

1. Concentrated sulfuric acid and concentrated nitric acid **[1]**
2. Catalyst **[1]**
3. a) Moderate heat / 50°C **[1]**
       reflux **[1]**
       $AlCl_3$ **[1]**
    b) $C_6H_6 + HCOCl \rightarrow C_6H_5CHO + HCl$ **[1]**
    c) Nucleophilic substitution **[1]**
4. a) $(3 \times 612) + (3 \times 348) + (6 \times 412) = 5352$ **[1]**
       $(6 \times 715) + (6 \times 218) = 5598$ **[1]**
       $\Delta H_2 = M2 - M1 - 83 = +163$ kJ mol⁻¹ **[1]**
    b) p delocalised electrons provide increased stability **[1]**

Page 125

1. Cyclohexylamine **[1]**
2. Nucleophilic substitution **[1]**
3. a) Concentrated H_2SO_4 **[1]**
       Concentrated HNO_3 **[1]**
    b) Nitrobenzene **[1]**
    c) Nucleophilic **[1]**
       addition–elimination **[1]**

4. a)

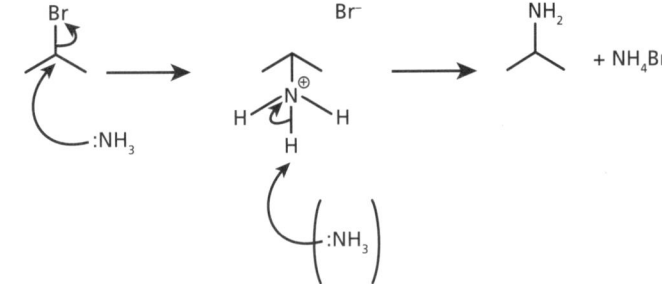

For curly arrow from lone pair on N to C [1]
For curly arrow from bond to Br [1]
For structure of intermediate [1]
For loss of H$^+$ [1]

b) Both the product of the reaction and ammonia are bases. [1]
They can both accept a proton on the lone pair of electrons on the nitrogen. [1]
Propan-2-amine [1] is a stronger base than ammonia. [1]

Page 127

1. Poly(but-1-ene) [1]
2. Terylene [1]
3. a) Polyester [1]
   b)

   $$-\overset{\overset{O}{\|}}{C}-\overset{\overset{O}{\|}}{C}-O-CH_2-CH_2-CH_2-O-$$ [1]

4. a) $x = 5$ [1]
   $y = 9$ [1]
   b) Van der Waals/London forces/induced dipole-dipole forces and hydrogen bonds [1]
   c) Amide bonds can undergo hydrolysis [1]
   by water/conditions found in the environment chemically breaking down the polymer back into the monomers/dicarboxylic acids and diamines [1]

Page 129

1. Hydrogen bonds [1]
2. $H_3N^+-CH-COO^-$
   $\quad\quad\quad |$
   $\quad\quad H_2C-SH$ [1]

3. a) The difference in the balance between solubility in solvent/mobile phase and attraction to/retention on stationary phase. [1]
   b) i) Nihydrin / Iodine [1]
      ii) Amino acids are colourless so to make them visible [1]
   c) $R_f = \dfrac{\text{distance travelled by substance}}{\text{distance travelled by solvent}}$ [1]
   $= 0.34 \times 80 = 27.2$ mm [1]

4. a) Secondary [1]
   b) Hydrogen bonds [1]
   between the lone pair of electrons on the O and the δ+H. [1]

This arises because nitrogen and oxygen are very electronegative. [1]
C=O and N–H are polar. [1]

Page 131

1. Enzymes contain a stereospecific active site that can only bond to one enantiomeric form of a substrate. [1]
2. $CH_3-CH-COOH$
   $\quad\quad\quad |$
   $\quad\quad\quad OH$ [1]
3. a) Protein [1]
   A (biological) catalyst / speeds up biochemical reactions. [1]
   b) Active site / the region where the substrate binds has a unique shape [1] complementary to the shape of only one specific substrate molecule. [1]
   c) A molecule with a similar shape to the substrate that binds to the enzyme molecule [1]
   Prevents active site from binding to substrate [1] and slows down / stops enzyme's catalytic activity. [1]
4. Enzyme has an active site, [1]
   which only fits one enantiomer. [1]
   The G isomer has the correct stereo chemistry to bind / enzyme has the complementary shape to the G isomer. [1]
   The enzyme-inhibitor complex hydrolyses at a faster rate than the F isomer, leaving only isomer G. [1]

Page 133

1. Cytosine and guanine [1]
2. Replacement of two Cl$^-$ ligands with two N on two guanine bases [1]
3. a) Prevents DNA replication [1]
   leading to the death of rapidly dividing cells. [1]
   b) $[Pt(NH_3)_2Cl_2] + H_2O \rightarrow [Pt(NH_3)_2Cl(H_2O)]^+ + Cl^-$
   Correct formula of the complex ion [1]
   Correct balanced equation [1]
   c)

   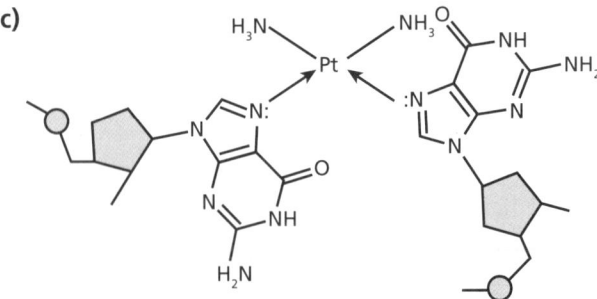

   For covalent bonds from Nitrogen to Pt complex [1]
   For arrow on covalent bonds or lone pair of electrons on nitrogen [1]

4. a) X = base **[1]**

Y = phosphate **[1]**

b) Polymer of nucleotides **[1]**

linked by covalent bonds **[1]**

between the phosphate group of one nucleotide and the 2-deoxyribose of another nucleotide. **[1]**

c) Two complementary strands **[1]**

(forming a) double helix **[1]**

due to the hydrogen bonding between base pairs on the two complementary strands. **[1]**

Page 135

1. But-1-ene **[1]**

2. NaCN then dilute HCl **[1]**

3. a) Step 1: Concentrated H_2SO_4 / Concentrated H_3PO_4 **[1]**

Step 2: Bromine water / Br_2 **[1]**

Step 3: Strong alkali / NaOH / KOH **[1]**

b) Any suggestion giving the idea of: Less energy used / Better yield / Reduces practical losses **[1]**

And any suggestion giving the idea of: Less waste / Less pollution / Maximises the use of raw materials to make useful products / Saves resources **[1]**

4. a) Step 1: addition **[1]**

Step 2: substitution **[1]**

a) Step 1: Cl_2 **[1]**

Step 2: KCN **[1]**

Step 3: H_2 / Ni **or** $LiAlH_4$ **or** Na / C_2H_5OH **[1]**

Page 137

1. Tetrachloromethane **[1]**

2. Its ^1H NMR spectrum has two peaks with an integration ratio of 2 : 3 **[1]**

3. a) TMS is inert or non-toxic or volatile / easily removed **[1]**

It gives a single (intense) peak **[1]**

b)

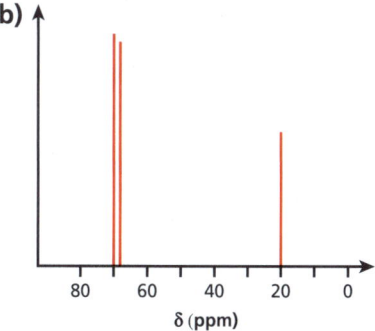

Three peaks **[1]**

Two peaks around 70 ppm and one peak around 20 ppm **[1]**

4.

$$CH_3 - \overset{\underset{\displaystyle O}{\|}}{C} -$$ **[1]**

$$-OH$$ **[1]**

$$-CH_2 - CH_2 -$$ **[1]**

$$CH_3 - \overset{\underset{\displaystyle O}{\|}}{C} - CH_2 - CH_2 - OH$$ **[1]**

Page 139

1. The balance between solubility in the moving phase and retention by the stationary phase **[1]**

2. Gas chromatography **[1]**

3. a) Silica gel = stationary phase **and** hexane = mobile phase **[1]**

b) Cyclohexene is less polar than cyclohexanol / cyclohexanol is more polar than cyclohexene/ cyclohexene is non-polar and cyclohexanol is polar **[1]**

Cyclohexene has a greater affinity/attraction for the mobile phase/hexane / cyclohexanol has a greater affinity/attraction for the stationary phase/silica **OR** cyclohexene is more soluble in the mobile phase/hexane or converse for cyclohexanol **[1]**

4. a) nihydrin **[1]**

b) $R_f = \frac{53}{98}$ **[1]** = 0.54 **[1]**

PRACTICE PAPERS

Note: 'ecf.' stands for 'error carried forward'. If you get the wrong answer for part of a question and carry this mistake through to the next part of the question, 'allow ecf.' indicates that you can still get full marks if you do this part correctly.

Paper 1

Question number	Marking guidance	Mark	Comment and tips on examination technique
01.1	$1s^2 2s^2 2p^6 3s^2 3p^6 4s^2 3d^{10}$ (allow $1s^2 2s^2 2p^6 3s^2 3p^6 3d^{10} 4s^2$)	1	*Zinc has atomic number 30, so fill up orbitals in sequence. Always double-check where your electron configuration would place your element on the Periodic Table: here it would be at the end of the d-block, period 4.*
01.2	Zinc does not form an ion with incomplete d-orbitals because its ion is Zn^{2+} and it loses its 4s electrons first	1 1	*'d-block' and 'transition' are subtly different: learn the definition of a transition element and remember that, although 4s fills before 3d, it usually empties before 3d as well.*
01.3	Octahedral and the diagram should be: $\left[\begin{array}{c} OH_2 \\ H_3N \cdots Cu \cdots NH_3 \\ H_3N \quad OH_2 \quad NH_3 \end{array}\right]^{2+} \left[\begin{array}{c} OH_2 \\ H_3N \cdots Cu \cdots NH_3 \\ H_3N \quad OH_2 \quad NH_3 \end{array}\right]^{2+}$	2	*Six ligands (four ammonia ligands and two water ligands) around a central copper ion. The overall shape is described as 'octahedral' because, in three dimensions, the ion has eight sides.*
01.4	$[Co(H_2O)_6]^{2+}(aq) + 4Cl^-(aq) \rightarrow [CoCl_4]^{2-}(aq) + 4H_2O(l)$	1	*This is a ligand-substitution reaction, with chloride ions forming a more stable complex. Don't forget that the fact that the chloride ligand has a negative charge changes the overall charge on the complex ion.*
02.1	Enthalpy change when **1 mole of gaseous atoms become gaseous ions, carrying a single positive charge**, under standard conditions (allow ΔH^{θ}: $M_{(g)} \rightarrow M^+_{(g)} + e^-$)	 1	*The definition should be learned and must include 1 mole and the gaseous state. The equation saves time and space because it states exactly the same as the worded definition.*
02.2	Trend: is a general increase Na–Ar Explanation: the **positive charge of the nucleus increases** by 1 with each successive element, making the electron more difficult to remove	1 1	*Although the trend is not perfect, there is clearly only 1 mark available for stating it, so a general trend will do. Electrons are in the same energy level, so the charge on the nucleus becomes the most important factor in determining ionisation energy.*

Question number	Marking guidance	Mark	Comment and tips on examination technique
02.3	\n\nThe graph should have one point showing that **the first electron requires significantly less** energy to remove than the next eight\n\nThe graph should have two points showing that the **last two electrons require significantly more** energy to remove than the preceding eight	1\n\n1	*Have the 2, 8, 1, arrangement in mind as you put points onto the graph but remember that, when removing electrons, it becomes 1, 8, 2! The big increases in energy happen as you move from one energy level (or shell) into the next.*
02.4	**Ionisation** (by bombarding with electrons)/ impact ionisation or electrospray ionisation turns the atoms into ions with a positive charge\n\n**Acceleration** (by negative plates) gives all ions the same kinetic energy\n\n**Ion drift** separates the ions according to their different masses because lighter ions will travel faster than the heavier ones\n\n**Detection** allows the time of flight to be calculated	1\n\n1\n\n1\n\n1	*These are the points that have to be learned about the time of flight technique. Kinetic energy is mass × velocity, so particles with different masses, but the same kinetic energy, will have different velocities, so different times of flight.*
02.5	75% ^{85}Rb\n\n25% ^{87}Rb	1	*You need to solve this equation:*\n\n*$85.5 = 85(x) + 87(100 - x)$*\n\n*Then $85x + 87(100 - x) = 8550$*\n\n*$85x + 8700 - 87x = 8550$*\n\n*$-2x = -150$*\n\n*So $x = 75\%$*
03.1	$MgSO_4(aq) + BaCl_2(aq) \rightarrow BaSO_4(s) + MgCl_2(aq)$\n\n(mark not awarded if state symbols not included)	1	*Barium and magnesium are both in Group 2, so both have the same type of formula for their compounds. State symbols are essential here, as the precipitation is the important feature.*
03.2	Moles of barium chloride = 0.10 moles\n\nMaximum possible number of moles of magnesium sulfate is 10/120 moles = 0.0833 moles\n\nStoichiometry is 1:1, so barium chloride is in excess	1\n\n1\n\n1	*Easy to calculate the number of moles of barium chloride, but tricky to have to realise that you have to calculate the number of moles of magnesium sulfate without the water, as a maximum. With water present, the number of moles will be even less.*

Question number	Marking guidance	Mark	Comment and tips on examination technique
03.3	Number of moles of barium sulfate = 9.47/233 = 0.04 moles (number of moles of magnesium sulfate = 0.04 moles) 0.04 × 120 = 4.88 g of magnesium sulfate in 10 g, so 5.12 g water Moles water = 5.12/18 = 0.284 moles so ratio is 0.284/0.04 = 1:7.1 and x = 7 (allow all 3 marks if x = 7)	1 1 1	The quantity you know is the number of moles of barium chloride, so work that out first and, because it is 1:1, this is the number of moles of magnesium sulfate. Multiply by the formula mass of magnesium sulfate to work out the mass of water in 10 g and convert that into moles of water. What you need, to work out x, is the ratio of moles of water to moles of magnesium sulfate. Always show your working-out as you go, so an incorrect value for x can still gain you marks.
03.4	Based on mass change on converting to anhydrous magnesium sulfate: Measure mass of hydrated magnesium sulfate Heat, in a crucible, until no further loss in mass Measure mass of anhydrous magnesium sulfate Work out the mole ratio of water lost to moles of anhydrous magnesium sulfate	1 1 1 1	This is the more usual method to work out the number of moles of water per mole of hydrated salt. The key to gaining all available marks is to remember to weigh the solid before and after heating, and to remember to mention that the masses will need converting to moles, to work out the mole ratio.
04.1	$K_c = \dfrac{[SO_3]^2}{[SO_2]^2[O_2]}$ (allow 1 mark if concentrations not raised to correct power) $0.94/2.31 \times 10^{-25} = 4.07 \times 10^{24}$ (allow ecf. from incorrect K_c expression) Units are $mol^{-1} \, dm^3$	2 1 1	Always products on the top and reactants on the bottom, then the powers to raise concentrations to are the numbers of moles in the equation. Show how you worked out the numerical value of K_c. Units for SO_3 and SO_2 cancel out, leaving you with 1/mol dm^{-3}.
04.2	C	1	The reaction is exothermic in the forward direction and will therefore be favoured by a decrease in temperature.
04.3	C	1	Although pressure affects the position of equilibrium, equilibrium is only reached when the forward and reverse reactions proceed at the same rate.
04.4	Yield is unchanged as the position of equilibrium remains the same. Rate increases/rate of the forward reaction is increased by the same amount as the rate of the reverse reaction so equilibrium is reached faster.	1 1	Catalysts affect rate of reaction and not position of equilibrium.
04.5	Acid rain $H_2O + SO_3 \rightarrow H_2SO_4$ or the equivalent, in words, or for sulfurous acid, to show how the acid is formed	1 1	Two marks for this question tell you that 'acid rain', alone, will not be all that you need to write in order to achieve full marks.

Question number	Marking guidance	Mark	Comment and tips on examination technique
05.1	Ability to attract a pair of electrons in a covalent bond (or words to that effect)	1	*Learn definitions: this definition has a 'pair of electrons'.*
05.2	Trend: electronegativity decreases going down the group, from F–I	1	*Although the nucleus becomes more positively charged as you go down the group, the increase in shielding has much more effect.*
	Explanation: atoms have more electrons shielding the nucleus, reducing its attraction	1	
05.3 (i)	$Cl_2(g) + 2Br^-(aq) \rightarrow Br_2(aq) + 2Cl^-(aq)$ (ignore state symbols)	1	*Don't forget halogens are diatomic elements!*
05.3 (ii)	Cl_2 (allow Cl, or chlorine)	1	*The oxidising agent is the species that is reduced in the reaction: Cl changes from 0 to –1, which is reduction.*
05.4	$Br_{2(l)} \rightarrow Br_{2(g)}$	1	*Boiling, or evaporating, only changes the state symbols, not the formula.*
05.5	Fluorine molecules have fewer electrons than bromine molecules	1	*Questions comparing boiling points, or melting points, are usually about intermolecular forces. Make sure you name the force between the molecules and explain the difference in strength.*
	The intermolecular forces – Van der Waals/dispersion/London forces/induced dipole-dipole forces – are weaker in fluorine	1	
05.6	Sublimation or subliming	1	*A correct term to be aware of, describing a direct change from solid to gas, without passing through a liquid phase.*
06.1	Na and Mg basic (Al amphoteric) Si, P, S and Cl acidic (allow 1 mark for 'metals basic, non-metals acidic')	1	*You must mention each of the elements, separately, as the question asks you to do so. You need to be aware that all s- and d-block elements have basic oxides, p-block metals are usually amphoteric and the remaining non-metal elements have acidic oxides.*
		1	
06.2	$Na_2O(s) + H_2O(l) \rightarrow 2NaOH(aq)$ (state symbols not necessary for the mark)	1	*This is a reaction and not just a dissolving of the oxide.*
06.3	$P_4O_{10}(s) + 6H_2O(l) \rightarrow 4H_3PO_4(aq)$ (state symbols not necessary for the mark)	1	*If you know the formula for the phosphoric acid that is formed, this is just a balancing exercise.*

Question number	Marking guidance	Mark	Comment and tips on examination technique
06.4	6 marks to be awarded for including the following points.		The reason for choosing three examples is that the oxides of the Period 3 elements show three different types of structure. Metallic oxides are generally ionic and non-metallic oxides are generally simple covalent molecules. Silicon dioxide, being giant covalent, is an unusual structure.
	• Na_2O: giant ionic lattice	1	
	• Strong electrostatic attraction between oppositely charged ions	1	
	• SO_2: simple molecular structures	1	
	• Weak intermolecular forces/weak induced dipole-dipole forces between molecules	1	
	• SiO_2: giant covalent structure	1	
	• Strong covalent bonds between atoms	1	
06.5	$CaO(s) + SiO_2(S) \rightarrow CaSiO_3(s)$ (state symbols not necessary for the mark)	1	The silicate ion is formed by the reaction of silicon dioxide with O^{2-}, so it must also have a 2- charge.
07.1	$NaOH(aq) + HCl(aq) \rightarrow H_2O(l) + NaCl(aq)$ (state symbols not required for the mark)	1	Acid reacts with base to produce a salt and water, which is the basis for neutralisation.
07.2	Fully ionises in solution.	1	Learn the definition of strong and weak, in relation to acids and bases.
07.3	$\dfrac{28.75 \times 0.1}{1000} = 2.875 \times 10^{-3}$ moles (allow both marks for the correct answer, without calculation)	2	Although you can achieve both marks for the correct answer, it is always best to include your calculation method in case, for example, you miscalculate the mean titre value.
07.4	Moles of sodium hydroxide = 2.875×10^{-3} $C = \dfrac{n}{v} = \dfrac{28.75 \times 10^{-3}}{0.025} = 0.115$ mol dm^{-3} (allow all 3 marks if the answer is correct, with units)	1 2	The most common mistake in this sort of calculation is to forget that the volume of the alkali being titrated is a 25.0 cm^3 portion of the original 250 cm^3. As volumes are in cm^3 and concentration is always in dm^{-3}, the volume must be divided by 1000.
07.5	Moles of sodium hydroxide in 250 cm^3 solution = 2.875×10^{-2} Mass of sodium hydroxide $= 2.875 \times 10^{-2} \times 40 = 1.15$ g Two pellets dissolved, so 0.575 g (allow all 3 marks if the answer is correct, with units)	1 1 1	As in 7.4, remember to work back to the original solution, which was made up to 250 cm^3. It will contain ten times as many moles of sodium hydroxide as are in the titrated portion. Many candidates would fail to notice that two pellets were originally dissolved.

Question number	Marking guidance	Mark	Comment and tips on examination technique
08.1	$2CO_{2(g)}$ + $3H_2O_{(1)}$ (enthalpy cycle diagram) -1560.0 kJ mol^{-1}, -1411.0 kJ mol^{-1}, -286.0 kJ mol^{-1}, $C_2H_{6(g)} \rightarrow C_2H_{4(g)}$ + $H_{2(g)}$ $\Delta H = -1560.0 + 1411.0 + 286.0$ $= +137.0$ kJ mol^{-1}	1 1	If a question suggests you include an enthalpy cycle diagram, and gives you space to include one, it would be unwise not to. In this example, the appropriate cycle would be a Hess' law triangle, with the common combustion products as the third corner, these being carbon dioxide and water. Because ΔH_c values given are per mole of the molecule being burned, the numbers of moles of the products do not need to balance for the mark.
08.2	$220 + 131 - 230$ $= 121$ J K^{-1} mol^{-1}	1	Standard molar entropies can be used to calculate entropy change in a reaction: $\Delta S^\theta = \sum S^\theta_{products} - \sum S^\theta_{reactants}$ Don't forget to include the units!
08.3	Entropy of the system increases as there is more disorder There are more particles in the products than in the reactants so there are more possible arrangements and ways to distribute energy among those particles	1 1	Endothermic reactions decrease the entropy of the surroundings because they decrease the temperature and ΔG and there are fewer ways to distribute the energy.
08.4	$\Delta G = \Delta H - T\Delta S_{system}$ $= 137\,000 - (298 \times 121)$ $= +100\,942$ J mol^{-1} $= +101.0$ kJ mol^{-1} (allow $+100.942$ kJ mol^{-1})	1 1	A common mistake is to forget that entropy changes are calculated in joules, whereas enthalpy changes are generally in kilojoules. Combining both in a calculation requires you to change the units so that both have the same units.
08.5	Feasibility just gives information that a reaction can happen not how long it will take to happen.	1	The reaction is feasible when $T\Delta S_{system}$ becomes larger than ΔH, giving ΔG a negative value. Reaction is only feasible at high temperatures.
08.6	A	1	The reaction is endothermic and so more energy is needed to break the bonds in the reactants than is released when the product bonds form.
09.1	Resists changes in pH On the addition of small amounts of acid, alkali or on dilution.	1 1	Buffers do not maintain a constant pH, so the mark is not awarded for a statement that implies no change in pH.
09.2	$C_2H_5COOH \rightleftharpoons C_2H_5COO^- + H^+$	1	The components of a buffer solution are an acid and its conjugate base, in this case sodium propanoate, produced by reacting propanoic acid with sodium hydroxide.

Question number	Marking guidance	Mark	Comment and tips on examination technique
09.3	(Solution contains both an acid and its conjugate base) Propanoic acid neutralises added base/ $C_2H_5COOH + OH^- \rightarrow C_2H_5COO^- + H_2O$ Propanoate neutralises added acid/ $C_2H_5COO^- + H^+ \rightarrow C_2H_5COOH$	1 1	*Normally an acid and a base cannot be present in the same solution, as they would react together, but a weak acid does not react with its conjugate base, leaving both able to neutralise small amounts of added acid or base.*
09.4	$pH = -\log \sqrt{K_a \times \text{concentration}}$ $= -\log \sqrt{1.3 \times 10^{-6}}$ $= 2.94$ (award both marks for correct answer without working, or calculated using a different method)	1 1	*Propanoic acid is a weak acid, so we arrive at this formula by making the assumption that the original concentration has not been significantly altered by the ionisation. After a calculation of this sort you should always check that the answer is realistic: weak acid pH will be between 1 and 7.*
09.5	Initial moles propanoic acid = 0.01 Initial moles sodium hydroxide = 0.005 (both needed for mark) After reacting, the buffer contains 0.005 mol of propanoic acid and 0.005 mol of sodium propanoate, in 200 cm³ Propanoic acid concentration = 0.025 mol dm⁻³ Sodium propanoate concentration = 0.025 mol dm⁻³ $pH = pK_a = 4.89$ (allow all marks for the correct answer worked out by an alternative method, but must include the concentrations)	1 1 1	*In making this buffer, the strong base will have reacted with half of the acid, leaving the other half as unreacted acid and making the same number of moles of conjugate base as there were moles of the strong base.* *Most questions about buffers involve mixtures that have been made such that the concentrations of acid and base are equal. Whenever this is the case, remember $pH = pK_a$.*
10.1 a)	i) $CH_3OH + H_2O \rightarrow CO_2 + 6 H^+ + 6 e^-$	1	
	ii) $O_2 + 4 H^+ + 4 e^- \rightarrow 2 H_2O$	1	
b)	$CH_3OH(l) + 1 O_2(g) \rightarrow CO_2(g) + 2 H_2O(l)$	1	
c)	Continuous supply of reactants/chemicals /fuel	1	
d)	Constantly add reactants	1	
	Keep concentration of reactants constant	1	
e)	Methanol (is liquid so) can be stored easily or transported easily	1	
	More energy can be produced from 1 cm³ of methanol (liquid) than from 1 cm³ of hydrogen (gas)	1	
f)	Same overall reaction	1	
g)	CO_2 released by fuel cell	1	
	equals (atmospheric) CO_2 taken up in photosynthesis		

Question number	Marking guidance	Mark	Comment and tips on examination technique
10.2	 d.c. power supply, labelled, (allow battery symbol)	1	Because the products at the electrodes are gases, the electrodes need to point upwards, with something to collect the gases in above each. Remember, electrolysis needs a d.c. supply and the anode is the positive electrode, whilst the cathode is the negative electrode.
	Correctly labelled (inert) cathode, connected to the negative terminal	1	
	Correctly labelled (inert) anode, connected to the positive terminal	1	
	Inverted test tubes, or equivalent, to collect gaseous products	1	
10.3	Anode: $4OH^- \rightarrow O_2 + 2H_2O + 4e^-$ Cathode: $2H^+ + 2e^- \rightarrow H_2$ (ignore state symbols)	1 1	When a solution is electrolysed, it is the ions present that are discharged at the electrodes.
10.4	concentration increases water is electrolysed from the solution hydroxide ions from water more readily discharged than sulfate ions	1 1 1	Sulfuric acid provides hydrogen ions to discharge at the cathode, and the hydroxide ion discharges at the anode more readily than the sulfate ion. The solution of sulfuric acid becomes more concentrated as it loses water.

Paper 2

01.1	Order: first order Reason: concentration of bromobutane trebled, rate trebles also	1 1	If the change in concentration is matched by the same change in rate, it is a first-order reaction.
01.2	Order: first order Reason: concentration of hydroxide doubles, rate doubles also	1 1	This deduction comes from comparing experiments 2 and 3: changing bromobutane concentration should decrease the rate to 2/3 of the original and then doubling the hydroxide ion concentration then doubles that rate.
01.3	$r = k[C_4H_9Br][OH^-]$	1	
01.4	$k = 0.08$ $dm^3\ mol^{-1}\ s^{-1}$	1 1	$4.0 \times 10^{-4}/(0.05 \times 0.1)$ Units for k turn $mol^2\ dm^{-6}$ into $mol\ dm^{-3}\ s^{-1}$.

Question number	Marking guidance	Mark	Comment and tips on examination technique
01.5	Change: k increases Explanation: at higher temperature there are more collisions At higher temperature a greater proportion of collisions have energy higher than E_A	1 1 1	The value of k is an indication of the rate of the reaction and it always increases with increasing temperature. Remember that 'more collisions' is a separate statement from 'more successful collisions'.
02.1	$Cl_2 \rightarrow 2Cl\bullet$ / 'homolytic fission of Cl–Cl bond' $Cl\bullet + C_2H_6 \rightarrow HCl + C_2H_5$ $C_2H_5 + Cl_2 \rightarrow C_2H_5Cl + Cl\bullet$ Free radical substitution	1 1 1 1	Free radical substitution reactions of alkanes need ultraviolet light to break the bond in Cl_2. The reaction then proceeds as a chain reaction. Remember to use the alkane mentioned in the question: it isn't always methane. Termination step isn't needed to show how chloroethane is formed.
02.2	Provides energy to break (covalent) bond in chlorine / Cl_2 or to form chlorine free radicals.	1	Homolytic means that, as the bond breaks, both product species have one of the electrons.
02.3	CFCs or free radicals **catalyse** the reaction	1	This is why CFCs are so damaging to the ozone layer.
02.4	$Cl\bullet + O_3 \rightarrow ClO\bullet + O_2$ $ClO\bullet + O_2 \rightarrow Cl\bullet + O_2$	2	This sequence needs to be learned because it is not easy to work out the mechanism, even when the reactants and products are known.
02.5	$Cl\bullet$ is regenerated and causes a chain reaction in the decomposition of ozone.	2	A 'chain reaction' means one that keeps going by producing one of the reactant species.
02.6	Norflane will not contribute to ozone depletion but halon could C–F bonds have higher bond energy / stronger covalent bonds than C–Cl C–F not broken by ultraviolet light but C–Cl is No halogen free radicals produced by norfluorane to break down ozone	1 1 1 1	CFC replacements are effective because it is specifically the C–Cl bond that is broken, creating $Cl\bullet$ free radicals in the upper atmosphere. Compounds containing C–F bonds are still 'greenhouse gases', although their impact on global warming is small compared to that of carbon dioxide.
03.1	 Correctly displayed –NH_2 (amine) group present Correctly displayed –COOH (carboxylic acid) group present	1 1 1	Despite the molecular formula giving no clue, the molecule is an amino acid and you are required to show the presence of amine and carboxylic acid groups.
03.2		1	Allow $(CH_3)_3{}^+N - CH_2 - COO$ (Br⁻)

Question number	Marking guidance	Mark	Comment and tips on examination technique
03.3	 Two enantiomers must be drawn, using 3D representation of bonding around the chiral carbon (allow 1 mark for correctly identifying the chiral carbon atom)	2	To make sure that your diagram shows two different enantiomers of the molecule, you should use the mirror-line techniques and make sure that the two versions you draw are mirror images. It is always a good idea to mark the chiral carbon with an asterisk.
03.4	An ion that has a positive and a negative charge (on different parts of the structure)	2	All amino acids, polypeptides and proteins exist as zwitterions in aqueous solution.
03.5 (i) (ii)	Either $^+H_3NC(CH_3)COO^- + H^+ \rightarrow \,^+H_3NC(CH_3)COOH$ Or $H_2NC(CH_3)COOH + H^+ \rightarrow \,^+H_3NC(CH_3)COOH$ Either $^+H_3NC(CH_3)COO^- + OH^- \rightarrow$ $H_2NC(CH_3)COO^- + H_2O$ Or $H_2NC(CH_3)COOH + OH^- \rightarrow$ $H_2NC(CH_3)COO^-$ $+ H_2O$	1 1	The amino acid zwitterion reacts with the proton from hydrochloric acid and the chloride ion is a spectator ion. If you include the chloride ion, it should be next to the $-NH_3^+$ group in the product. The amino acid zwitterion reacts with the hydroxide ion from sodium hydroxide and the sodium ion is a spectator ion. If you include the sodium ion, it should be next to the $-COO^-$ group in the product.
03.6	 GLY-ALA correctly drawn ALA-GLY correctly drawn	1 1	A dipeptide, formed by joining two amino acids, always has two possible arrangements, depending on which amino acid becomes the N–terminal and which becomes the C–terminal.
03.7	Paper chromatography/thin layer chromatography	1	Chromatography separates and identifies the components of a mixture. Thin layer chromatography tends to give more accurate results, especially when calculating R_f values.
04.1	Methyl benzoate	1	
04.2	Moles of benzoic acid $= 10.5/ 122 = 0.086$ mol 1:1 so theoretical yield is $0.086 \times 136 = 11.7$ g $\frac{9.98}{11.70} \times 100\% = 85.3\%$	1 1 1	Make sure that each step of the calculation is clearly written. So, if you make a slight arithmetic error, you will still get the working out marks.
04.3	$HNO_3 + H_2SO_4 \rightarrow NO_2^+ + HSO_3^- + H_2O$ nitronium ion Concentrated sulfuric acid is a stronger acid than nitric acid, so it protonates the nitric acid. The protonated nitric acid loses a molecule of water, forming the nitronium ion which is the electrophile.	1 1 1 1	A common mistake is to confuse the nitronium ion with the nitrate ion, NO_3^-, but this would not be an electrophile.

Question number	Marking guidance	Mark	Comment and tips on examination technique
04.4	Curly arrow from ring to NO_2^+ Intermediate with delocalisation over five carbons and the sixth bonded to H and NO_2 (positive charge inside the ring) Curly arrow from C–H bond into the ring Correct formula of nitrated product (allow nitration at any C_2, C_3, or C_4)	1 1 1 1	Remember curly arrows go from where electrons are to where they are not. Ring has delocalised electrons, nitronium ions lacks electrons. The intermediate only has the electron delocalisation over the other five carbons and you need a positive charge inside the ring. As the intermediate releases the hydrogen ion, the electrons return to the ring and delocalisation is complete again.
05.1	IR causes covalent bonds to vibrate (by bending or stretching), absorbing radiation The wavenumber/frequency of the absorbed radiation is specific to each functional group	1 1	This question asks you to demonstrate a general understanding of IR spectroscopy, rather than interpreting the IR spectrum given. It is easy to waste time by getting drawn into describing the spectrum before the question demands it.
05.2	1710 cm^{-1}	1	This is a characteristic peak that a good candidate would already be aware of, but confirm by checking the Data Booklet.
05.3	Reagent: Tollen's Solution Expected result: No observable change/ remains colourless solution with ketone silver mirror/silver metal precipitated with aldehyde. OR Reagent: Fehling's reagent Expected results: Changes from blue solution to form a brick red precipitate formed if an aldehyde is present. No observable change/stays blue solution with ketone. OR Reagent: (Acidified) **Potassium dichromate** Expected results: Aldehydes change the orange solution into a green solution. Ketone no observable change/stays green	1 1 1 1 1 1 1 1 1	Brady's reagent gives the same result for both aldehydes and ketones.
05.4	Butanone (correct name **and** formula needed for each mark)	2	An aldehyde almost always has a ketone isomer. 3 marks should suggest only three possible isomers. The unbranched chains give one aldehyde and one ketone: note that butan-2-one and butan-3-one are the same molecule, so it needs no number.

Question number	Marking guidance	Mark	Comment and tips on examination technique
06.1	$C_2H_4(g) + Cl_2(g) \rightarrow C_2H_4Cl_2(l)$ (correct state symbols)	1 1	Addition of halogens to alkenes make a single halogenated product, with two halogen atoms on adjacent carbon atoms.
06.2	Dipoles on Cl_2 shown and curly arrow from double bond to the delta positive Cl Curly arrow from Cl–Cl bond to negative chlorine Intermediate has single C–C and positive charge located on three-bonded carbon Second Cl has negative charge and curly arrow from this to positive carbon	1 1 1 1	The common mistakes in answering questions about the mechanism of electrophilic addition are to put the curly arrows the wrong way round and to forget to mark dipoles and charges. It is one of the pair of electrons from the bond that move to the electrophile, not the other way round. They move because the electrophile has an induced dipole and a partial positive charge. The intermediate is a carbocation, which is attractive to the halide anion.
06.3	1,2-dichloroethane	1	This needs the numbers because there is a 1,1 isomer.
06.4	Correct Z isomer (first diagram, above) labelled Correct E isomer (second diagram, above) labelled (allow naming by cis-trans system) Name of type of isomerism: geometric (allow E/Z or cis-trans isomerism) **No free rotation** around the C=C double-bond (mark will not be awarded for C=C alone)	1 1 1 1	E/Z naming is not exactly the same as naming by 'cis' and 'trans' but, in this case, the cis isomer is 'Z' and the trans isomer is 'E'. It is the lack of free rotation that is particularly significant here and the mark would be lost if the answer given was just 'a C=C double-bond'.
06.5	1,2-dichloropropane would produce three peaks 2,2-dichloropropane would produce only one peak	1 1	There are three hydrogen environments in 1,2-dichloropropane, with hydrogen atoms on each of the three carbon atoms. 2,2-dichloropropane has hydrogen atoms on two of the carbon atoms but, because the molecule is symmetrical, the two end -CH_3 groups are equivalent.
07.1	Terylene	2	Terylene is a co-polymer so two repeat units means four of the original monomers have to be shown. 1 mark is for showing the correct monomer residues, which were named in the question. The second mark is for having them correctly linked and showing the four monomer residues.

Question number	Marking guidance	Mark	Comment and tips on examination technique
07.2	Poly(ethene) is formed from addition polymerisation and Terylene is condensation polymerisation To make polyethene/in addition polymerisation there is only one product To make Terylene/In condensation polymerisation there are two products the polymer and a small molecule/water	1 1 1 1 1	The molecule produced at each link, in this case, is water. Addition polymers only produce the polymer chain.
07.3	Terylene / polyesters and can hydrolyse Alkali produces OH^- ions which increases the rate of hydrolyse of Terylene Makes the bottle not durable / risk of it spilling its hazardous contents in storage	1 1 1	The clue is that the bottles have to contain alkali. Polythene is a suitably inert material, whereas Terylene will hydrolyse and produce the original monomers.
08.1	Stage 1: potassium dichromate/$K_2Cr_2O_7$ **and** sulfuric acid (allow 'acidified potassium dichromate') Stage 2: methanol sulfuric acid	1 1 1 1	To make an ester from an aldehyde, the aldehyde would first have to be oxidised to produce a carboxylic acid and then the carboxylic acid reacted with an alcohol in the presence of an acid catalyst.
08.2	M_r of methyl cinnamate = 162 100 g = 0.617 mol Plan to make 1.03 mol/166.7 g because of 60% yield Need 1.03 mol/152.4 g of the reaction intermediate Plan to make 1.71 mol/253.1 g of the reaction intermediate, because of 60% yield From 1.71 mol of cinnamaldehyde 1.71 × 132 = 225.7 g	1 1 1 1 1	This question can be worked through, using either moles or mass. As each stage has only a 60% yield, you need to plan to make 166.7% of the methyl cinnamate from the reaction intermediate and 166.7% of the reaction intermediate from the cinnamaldehyde.
08.3	A: benzaldehyde B: methanoic anhydride	1 1	Naming the benzaldehyde should be quite straightforward, but the methanoic anhydride relies on identifying the acid anhydride functional group. You have probably seen this in ethanoic anhydride, which is useful in producing esters.
08.4	**Any two from:** • Availability of raw materials • Toxicity of raw materials • Rate of the reaction • Atom economy of the reaction • Yield that can be achieved	 2	These are the important considerations, along with cost, which determine the viability of a synthesis.

Question number	Marking guidance	Mark	Comment and tips on examination technique
09.1	Tertiary	1	Count the number of carbon atoms bonded to the nitrogen atom: Seliginine has three and norepinephrine has only one.
09.2	**Structural similarity** suggests Seliginine binds to the active site of the enzyme Competitive inhibition	1 1	Competitive inhibition is likely when both the substrate and the inhibitor are able to bind to the active site of the enzyme. Because the active site has a specific shape, a competitive inhibitor usually resembles the substrate molecule in its structure.
09.3	A	1	The primary structure of a protein is the sequence of amino acids, joined together by covalent bonds.
09.4	B	1	The secondary structure of a protein includes an α-helix and β-pleated sheet, held in place by hydrogen bonds.
09.5	Proteins are made from amino acid residues **and** DNA is made from nucleotides (both needed for the mark) DNA is a double-stranded molecule **and** proteins are single-stranded (both needed for the mark)	1 1	When you write an answer to compare two things, you must mention both things in the comparison you make. If you answered this question with 'DNA is double stranded and made from nucleotides', you would not be awarded either mark. Beware of using the word 'polymer' to describe a protein or DNA, because the correct term is 'macromolecule'. Neither a protein or a DNA molecule has repeat subunits.

10.1	21.3% oxygen
	C: 72/100 × 150 = 108
	    108/12 = 9
	H: 6.7/100 × 150 = 10.05
	    10.05/1 = 10
	O: 21.3/100 × 150 = 32
	    32/16 = 2
	Molecular formula: $C_9H_{10}O_2$
		1
		
		1
		1
			The elemental analysis gives the percentage of composition of elements in the sample. As you are only given the percentages of two of the three elements, you must first work out the percentage composition of oxygen. Then the moles of each element, which is the same as the ratio of the elements in the substance. The empirical formula is then generated from the simplest whole number ratio of the elements in the substance. To generate the molecular formula, you need to calculate the mass of the empirical formula and work out how many empirical formulas are needed to make the relative molecular mass.

Question number	Marking guidance	Mark	Comment and tips on examination technique
10.2	 (allow displayed formulae)	1 1 1	*With six of the carbon atoms in a phenyl group (or benzene ring), the presence of only seven ^{13}C peaks confirms that the arrangement of groups around the ring must be symmetrical. One –COOH group at position 1, and one ethyl group, at position 4 is a possible structure.* *Two methyl groups, arranged 2,6 or 3,5 would only give six carbon environments, so the other isomers must either have one methyl group in position 4 and one CH_2 between –COOH and the ring, or two CH_2 groups between –COOH and the ring.* *The final clue is that with 3 marks available, you should expect to draw three isomers.*

Paper 3 – Section A

Question number	Marking guidance	Mark	Comment and tips on examination technique
01.1	M_r of sodium hydrogensulfate = 120 $120 \times 0.1 \times 1000/250$ = 3.0 g	 1 1	*You need to have noticed that only 250 cm³ of solution is required, so one quarter of 0.1 mol.*
01.2	Take the mass of the weighing boat after transfer from the mass of the weighing boat with the sodium hydrogensulfate This ensures that any residue left on the weighing boat is taken into account/on any transfer some substance will be left behind	1 1	*This is the only way of ensuring that only the added mass is measured.*
01.3	Percentage error = $\dfrac{error}{actual\ value} \times 100$ error of the balance = 0.01 Percentage error = $\dfrac{0.01}{3.00} \times 100 = 33.3$ %	1 1 1	*If value of 4.00 was used, the answer would be 25.0%*
01.4	Rinse **both the beaker and the glass rod**, transferring the rinsings to the volumetric flask	2	*The rinsings may contain some of the sodium hydrogensulfate.*
01.5	(i) Add distilled water dropwise near to the graduation mark Ensure that the bottom of the meniscus is in line with the graduation mark Ensure volume is read in eye line (ii) Mix solution	1 1 1 1	*Good titration technique, in both cases.*

Question number	Marking guidance	Mark	Comment and tips on examination technique
02.1	Zinc half-cell: $Zn_{(s)} \rightleftharpoons Zn^{2+}_{(aq)} + 2e^-$ Copper half-cell: $Cu_{(s)} \rightleftharpoons Cu^{2+}_{(aq)} + 2e^-$ (allow either to be written the other way around)	1 1	Each half-cell contains a metal/metal ion equilibrium.
02.2	As zinc ionises more readily than copper, **releasing electrons**, the metal electrode becomes more negatively charged	2	Don't be confused by the labels 'anode' and 'cathode', which, for a voltaic cell, are the opposite way around to the labelling in an electrolytic cell.
02.3	$+0.76 + 0.34 = +1.10$ V (298K, 101kPa, 1 moldm^{-3})	1 1	Change the sign on the cell that is reversed in relation to its standard electrode potential.
02.4	The zinc electrode becomes more negative	1	
02.5	Complete circuit / allow ions to move between half cells	1	Consider the equilibrium in the half-cell and apply Le Chatelier's principle. Reducing the ion concentration will cause greater ionisation of the metal electrode.
02.6	Reduction and oxidation both happen Copper gains electrons/decreases oxidation number (reduced) Zinc loses electrons/increases oxidation number (oxidised)	1 1 1	As well as stating the definition of 'redox', show the actual changes in oxidation numbers.
03.1	A (25.00 cm^3) pipette	1	For certain volumes a volumetric pipette is the best choice, as it is the most accurate.
03.2	$Mg + Cu^{2+} \rightarrow Mg^{2+} + Cu$	1	This is a displacement reaction, magnesium being the more reactive metal.
03.3	Moles copper sulfate $= \dfrac{25}{1000}$ $\times 0.5 = 0.0125$ mol $1 \times 25. 10^{-2}$ mol magnesium has reacted	1 1 1	There is a large excess of magnesium metal and only the same number of moles as there are of copper sulfate can have reacted.
03.4	Energy $= 25 \times 4.18 \times 5.5 = 574.75$ J $574.75/0.0125 = -45\,980$ J mol^{-1} (-45.98 kJ mol^{-1})	1 1	Remember the formula for calculating the energy change and then divide the answer by the number of moles of copper sulfate, as magnesium is in excess. Enthalpy change always has a sign, even though the energy change doesn't.
03.5	Source of error: heat lost to the surroundings Suggested improvement: increase insulation/add a lid/put insulation around the beaker	1 1	The question describes the experiment as having been carried out in a glass beaker. This is not a good piece of apparatus to use as a calorimeter.

Question number	Marking guidance	Mark	Comment and tips on examination technique
03.6	Effect: greater increase in temperature Explanation: more copper sulfate to react (in the same volume)	1 1	*Because of the excess magnesium, the number of moles reacting is the number of moles of copper sulfate.*
04.1	$C_2H_5OH + [O] \rightarrow CH_3CHO + H_2O$ (allow molecular formulae for ethanol/ethanal)	1	*This equation becomes a lot simpler, using [O]. Notice that ethanol loses two hydrogen atoms.*
04.2	Use of anti-bumping granules/glass beads Heat in a water bath/over a beaker of water	1 1	*Flammable mixtures should always be heated in a water bath or over water and the anti-bumping granules help break up large bubbles.*
04.3	Ice reduces the temperature so that the vapours condense	1 1	*Depending on the boiling and melting point will depend how you prepare to cool the vapours. You can use just ice, ice mixed with salts to get a lower temperature or even just cool air with an air condenser.*
04.4	Fehlings solution Solution turns from blue to give a red precipitate/copper precipitate Or: Acidified potassium dichromate from orange to green Or: Ammoniacal silver nitrate solution/Tollen's reagent Silver-mirror forms	1 1 1 1 1 1	*In choosing reagents for any chemical test, always describe what would constitute a positive result.* *Brady's reagent tests for a carbonyl compound, whereas Tollen's reagent is specifically to test for an aldehyde.*
04.5	Oxidation of ethanal Add sodium carbonate solution until there is no further effervescence Distil the mixture a second time	1 1 1	*It is always difficult to prepare pure aldehyde by oxidation of a primary alcohol because the same oxidising agent also oxidises the aldehyde.* *This is a useful technique because when the effervescence stops, the acid must have been neutralised.*
04.6	$n = \dfrac{m}{M_r}$ 4.6 g ethanol = 0.1 mol 3.8 g ethanal = 0.0864 mol 86.4%	1 1 1 1	*Remember how to calculate percentage yield:* $\dfrac{actual\ yield}{maximum\ theoretical\ yield} \times 100$
05.1	Burette: sodium hydroxide Pipette: ethanoic acid	1 1	*Items of glassware should be rinsed with the solution that will be put into them.*

Question number	Marking guidance	Mark	Comment and tips on examination technique
05.2	Need to choose an indicator where the end point is a good approximation for the equivalence point	1	Recognise that ethanoic acid is a weak acid and it is being titrated with a strong base, sodium hydroxide.
	The pH range of phenolphthalein is 8.2 to 10.0 / end point for phenolpthalein is slightly alkali	1	
	Sodium hydroxide is a strong alkali and ethanoic acid is a weak acid	1	
	Equivalence point for this titration will be slightly alkali/greater than 7	1	
	This indicator / phenolphthalein changes colour in the correct range	1	
05.3	Equivalence point is where the number of moles of acid equal the number of moles of alkali	1	Any titration has the greatest change of pH at the equivalence point.
	On a pH curve, it is the centre of the vertical section of the curve/sharp inflection of the curve	1	
05.4	The mixture is a **buffer solution**	1	The sodium hydroxide will react with the ethanoic acid to make sodium ethanoate. Buffering will be most effective when the concentrations of ethanoic acid and sodium ethanoate are approximately equal.
	The concentrations of ethanoic acid and sodium ethanoate are approximately equal.	1	

Paper 3 – Section B

Question number	Answer	Mark	Comment and tips on examination technique
06	A	1	Arenes react mainly by electrophilic substitution.
07	B	1	Reactivity is due to the energy and positions of electrons, not protons.
08	D	1	A quick calculation confirms that the calcium carbonate has only partially decomposed. The number of moles of carbon dioxide is the same as the number of moles of calcium oxide, which multiplied by $24.0 \ dm^3$ give the volume.
09	B	1	The only difference in properties between enantiomers is their rotation of the plane of plane-polarised light in opposite directions.
10	D	1	In energy level $n = 4$, there are a total of seven electrons. Halogens are Group 7.
11	C	1	Van der Waals < dipole-dipole < hydrogen bonding. The carboxylic acid has greater hydrogen bonding potential than the alcohol.
12	C	1	Remember: low pH means acidic. Non-metal oxides are acidic, but silicon dioxide is insoluble because of its giant covalent structure.
13	B	1	Multiplying the concentrations together gives $mol^3 \ dm^{-9}$. Rate is usually in $mol \ dm^{-3} \ s^{-1}$.
14	B	1	It has to be either B or C, because temperature does affect the position of equilibrium. Exothermic means increasing temperature favours reactants.

Question number	Answer	Mark	Comment and tips on examination technique
15	A	1	*Elimination is what is needed: high temperature and a strong base catalyst, with ethanol as the solvent, favours elimination.*
16	A	1	*Alcohols and aldehydes can be oxidised, ketones cannot be oxidised.*
17	C	1	*Octahedral is the only shape here with six corners.*
18	D	1	*Atomic radius decreases across a period because of increasing nuclear charge pulling outer-shell electrons closer.*
19	C	1	*You need to know $pV = nRT$ and how to use it to calculate the volume of gas under non-standard conditions.*
20	B	1	*Partial pressure is mole fraction × total pressure, so convert each number of moles to the mole fraction by dividing it by the total number of moles, 1.02. The partial pressures of NO_2 and NO both need to be squared in the equilibrium expression. You should be able to see why the units are atm too.*
21	A	1	*Sodium and potassium hydroxides are soluble. The solubility of Group 2 hydroxides increases as you go down the group.*
22	B	1	*Electron affinity is for 1 mole of atoms to become 1 mole of negative ions, all in the gaseous state.*
23	D	1	*A nucleophile is a 'lone pair donor' and always has a negative, or partial negative, charge.*
24	A	1	*pH of a weak acid is $-\log \sqrt{K_a \times \text{concentration}}$.*
25	B	1	*The positive enthalpy change means it is an endothermic reaction. $-T\Delta S$ has to be more negative than ΔH is positive for ΔG to be negative.*
26	B	1	*This is the weighted mean: add together A_r × % abundance for each isotope and divide by 100.*
27	A	1	*Just memory, but it is something you need to remember.*
28	D	1	*R_f is $\dfrac{\text{distance moved by spot}}{\text{distance moved by solvent}}$*
29	C	1	*Look for a big increase between the fifth and sixth ionisation energies.*
30	B	1	*This is the standard test for a halide: chlorides give a white precipitate.*
31	D	1	*Atom economy is the mass of useful product (the iron) divided by the total mass of products.* *112/244 = 45.9%*
32	B	1	*The chiral carbon needs four different groups bonded to it.*
33	B	1	*7.8 g of benzene is 0.1 mol. 0.1 mol of nitrobenzene is 12.3 g.*
34	C	1	*180° means a linear molecule, which rules out ammonia and the carbonate ion. Sulfur dioxide and carbon dioxide both have two double bonds but sulfur dioxide also has a lone pair of electrons, which decrease the bond angle.*
35	D	1	*Although the increase in oxidation number of the iron is +1, there are five iron ions per manganate VII ion in the balanced equation.*

THIS PAGE HAS DELIBERATELY BEEN LEFT BLANK

THIS PAGE HAS DELIBERATELY BEEN LEFT BLANK